4

D0395034

Through a Window

Books by Jane Goodall

Innocent Killers
(with Hugo van Lawick)

In the Shadow of Man

Grub the Bush Baby
(with Hugo van Lawick)

The Chimpanzees of Gombe:
Patterns of Behavior

Through a Window

*My Thirty Years
with the
Chimpanzees of Gombe*

<<<<<<<<<<<<<<<

Jane Goodall

Houghton Mifflin Company

BOSTON · 1990

Copyright © 1990 by Soko Publications Limited

ALL RIGHTS RESERVED

For information about permission to reproduce selections from
this book, write to Permissions, Houghton Mifflin Company,
2 Park Street, Boston, Massachusetts 02108.

Library of Congress Cataloging-in-Publication Data
Goodall, Jane, date.
Through a window : my thirty years with the chimpanzees of Gombe /
Jane Goodall.
p.   cm.
"First published in 1990 by George Weidenfeld & Nicolson . . .
London" — Galley t.p. verso.
ISBN 0-395-50081-8
1. Goodall, Jane, 1934–   . 2. Chimpanzees — Tanzania — Gombe
Stream National Park — Behavior 3. Zoologists — England — Biography.
I. Title.
QL31.G58A3   1990   90-36974
591'.092 — dc20   CIP
[B]

Printed in the United States of America

AGM 10 9 8 7 6 5 4 3 2 1

Dedication

To the chimpanzees of the world, those still living free in the wild and those held captive and enslaved by humans. For all that they have contributed to knowledge and understanding.

And to all those who have helped and who are helping in the fight to conserve the chimpanzees in Africa and to bring comfort and new hope to those in captivity.

And in memory of Derek.

Contents

Through a Window

Gombe

⋘⋘⋘⋘⋘

I ROLLED OVER and looked at the time — 5.44 a.m. Long years of early rising have led to an ability to wake just before the unpleasant clamour of an alarm clock. Soon I was sitting on the steps of my house looking out over Lake Tanganyika. The waning moon, in her last quarter, was suspended above the horizon, where the mountainous shoreline of Zaire fringed Lake Tanganyika. It was a still night, and the moon's path danced and sparkled towards me across the gently moving water. My breakfast — a banana and a cup of coffee from the thermos flask — was soon finished and, ten minutes later I was climbing the steep slope behind the house, my miniature binoculars and camera stuffed into my pockets along with notebook, pencil stubs, a handful of raisins for my lunch, and plastic bags in which to put everything should it rain. The faint light from the moon, shining on the dew-laden grass, enabled me to find my way without difficulty and presently I arrived at the place where, the evening before, I had watched eighteen chimpanzees settle down for the night. I sat to wait until they woke.

All around, the trees were still shrouded with the last mysteries of the night's dreaming. It was very quiet, utterly peaceful. The only sounds were the occasional chirp of a cricket, and the soft murmur where the lake caressed the shingle, way below. As I sat there I felt the expectant thrill that, for me, always precedes a day with the chimpanzees, a day roaming the forests and mountains of Gombe, a day for new discoveries, new insights.

Then came a sudden burst of song, the duet of a pair of robin chats, hauntingly beautiful. I realized that the intensity of light had changed: dawn had crept upon me unawares. The coming brightness of the sun had all but vanquished the silvery, indefinite illumination of its own radiance reflected by the moon. The chimpanzees still slept.

Five minutes later came a rustling of leaves above. I looked up and saw branches moving against the lightening sky. That was where Goblin, top-ranking male of the community, had made his nest. Then stillness again. He must have turned over, then settled down for a last snooze. Soon after this there was movement from another nest to my right, then from one behind me, further up the slope. Rustlings of leaves, the cracking of a little twig. The group was waking up. Peering through my binoculars into the tree where Fifi had made a nest for herself and her infant Flossi, I saw the silhouette of her foot. A moment later Fanni, her eight-year-old daughter, climbed up from her nest nearby and sat just above her mother, a small dark shape against the sky. Fifi's other two offspring, adult Freud and adolescent Frodo, had nested further up the slope.

Nine minutes after he had first moved, Goblin abruptly sat up and, almost at once, left his nest and began to leap wildly through the tree, vigorously swaying the branches. Instant pandemonium broke out. The chimpanzees closest to Goblin left their nests and rushed out of his way. Others sat up to watch, tense and ready for flight. The early morning peace was shattered by frenzied grunts and screams as Goblin's subordinates voiced their respect or fear. A few moments later, the arboreal part of his display over, Goblin leapt down and charged past me, slapping and stamping on the wet ground, rearing up and shaking the vegetation, picking up and hurling a rock, an old piece of wood, another rock. Then he sat, hair bristling, some fifteen feet away. He was breathing heavily. My own heart was beating fast. As he swung down, I had stood up and held onto a tree, praying that he would not pound on me as he sometimes does. But, to my relief, he had ignored me, and I sat down again.

With soft, panting grunts Goblin's young brother Gimble climbed down and came to greet the alpha or top-ranking male, touching his face with his lips. Then, as another adult male approached Goblin, Gimble moved hastily out of the way. This was my old friend Evered.

As he approached, with loud, submissive grunts, Goblin slowly raised one arm in salutation and Evered rushed forward. The two males embraced, grinning widely in the excitement of this morning reunion so that their teeth flashed white in the semi-darkness. For a few moments they groomed each other and then, calmed, Evered moved away and sat quietly nearby.

The only other adult who climbed down then was Fifi, with Flossi clinging to her belly. She avoided Goblin, but approached Evered, grunting softly, reached out her hand and touched his arm. Then she began to groom him. Flossi climbed into Evered's lap and looked up into his face. He glanced at her, groomed her head intently for a few moments, then turned to reciprocate Fifi's attentions. Flossi moved half-way towards where Goblin sat — but his hair was still bristling, and she thought better of it and, instead, climbed a tree near Fifi. Soon she began to play with Fanni, her sister.

Once again peace returned to the morning, though not the silence of dawn. Up in the trees the other chimpanzees of the group were moving about, getting ready for the new day. Some began to feed, and I heard the occasional soft thud as skins and seeds of figs were dropped to the ground. I sat, utterly content to be back at Gombe after an unusually long time away — almost three months of lectures, meetings, and lobbying in the USA and Europe. This would be my first day with the chimps and I planned to enjoy it to the full, just getting reacquainted with my old friends, taking pictures, getting my climbing legs back.

It was Evered who led off, thirty minutes later, twice pausing and looking back to make sure that Goblin was coming too. Fifi followed, Flossi perched on her back like a small jockey, Fanni close behind. Now the other chimps climbed down and wandered after us. Freud and Frodo, adult males Atlas and Beethoven, the magnificent adolescent Wilkie, and two females, Patti and Kidevu, with their infants. There were others, but they were travelling higher up the slope, and I didn't see them then. We headed north, parallel with the beach below, then plunged down into Kasakela Valley and, with frequent pauses for feeding, made our way up the opposite slope. The eastern sky grew bright, but not until 8.30 a.m. did the sun itself finally peep over the peaks of the rift escarpment. By this time we were high above the lake. The chimpanzees stopped and groomed for a while, enjoying the warmth of the morning sunshine.

About twenty minutes later there was a sudden outbreak of chimpanzee calls ahead — a mixture of pant-hoots, as we call the loud distance calls, and screams. I could hear the distinctive voice of the large, sterile female Gigi among a medley of females and youngsters. Goblin and Evered stopped grooming and all the chimps stared towards the sounds. Then, with Goblin now in the lead, most of the group moved off in that direction.

Fifi, however, stayed behind and continued to groom Fanni while Flossi played by herself, dangling from a low branch near her mother and elder sister. I decided to stay too, delighted that Frodo had moved on with the others for he so often pesters me. He wants me to play, and, because I will not, he becomes aggressive. At twelve years of age he is much stronger than I am, and this behaviour is dangerous. Once he stamped so hard on my head that my neck was nearly broken. And on another occasion he pushed me down a steep slope. I can only hope that, as he matures and leaves childhood behind him, he will grow out of these irritating habits.

I spent the rest of the morning wandering peacefully with Fifi and her daughters, moving from one food tree to the next. The chimps fed on several different kinds of fruit and once on some young shoots. For about forty-five minutes they pulled apart the leaves of low shrubs which had been rolled into tubes held closely by sticky threads, then munched on the caterpillars that wriggled inside. Once we passed another female — Gremlin and her new infant, little Galahad. Fanni and Flossi ran over to greet them, but Fifi barely glanced in their direction.

All the time we were climbing higher and higher. Presently, on an open grassy ridge we came upon another small group of chimps: the adult male Prof, his young brother Pax, and two rather shy females with their infants. They were feeding on the leaves of a massive *mbula* tree. There were a few quiet grunts of greeting as Fifi and her youngsters joined the group, then they also began to feed. Presently the others moved on, Fanni with them. But Fifi made herself a nest and stretched out for a midday siesta. Flossi stayed too, climbing about, swinging, amusing herself near her mother. And then she joined Fifi in her nest, lay close and suckled.

From where I sat, below Fifi, I could look out over the Kasakela Valley. Opposite, to the south, was the Peak. A surge of warm memories flooded through me as I saw it, a rounded shoulder perched above the long grassy ridge that separates Kasakela from the home valley, Kakombe. In the early days of the study at Gombe, in 1960 and 1961, I had spent day after day watching the chimpanzees, through my binoculars, from the superb vantage point. I had taken a little tin trunk up to the Peak, with a kettle, some coffee and sugar, and a blanket. Sometimes, when the chimps had slept nearby, I had stayed up there with them, wrapped in my blanket against the chill of the night air. Gradually I had pieced together something of their daily life, learned about their feeding habits and travel routes, and begun to understand their unique social structure — small groups joining to form larger ones, large groups splitting into smaller ones, single chimpanzees roaming, for a while, on their own.

From the Peak I had seen, for the first time, a chimpanzee eating meat: David Greybeard. I had watched him leap up into a tree clutching the carcass of an infant bushpig, which he shared with a female while the adult pigs charged about below. And only about a hundred yards from the Peak, on a never-to-be-forgotten day in October, 1960, I had watched David Greybeard, along with his close friend Goliath, fishing for termites with stems of grass. Thinking back to that far-off time I re-lived the thrill I had felt when I saw David reach out, pick a wide blade of grass and trim it carefully so that it could more easily be poked into the narrow passage in the termite mound. Not only was he using the grass as a tool — he was, by modifying it to suit a special purpose, actually showing the crude beginnings of tool-*making*. What excited telegrams I had sent off to Louis Leakey, that far-sighted genius who had instigated the research at Gombe. Humans were not, after all, the *only* tool-making animals. Nor were chimpanzees the placid vegetarians that people had supposed.

That was just after my mother, Vanne, had left to return to her other responsibilities in England. During her four-month stay she had made an invaluable contribution to the success of the project: she had set up a clinic — four poles and a thatched roof — where she had provided medicines to the local people, mostly fishermen and their families. Although her remedies had been simple — aspirin, Epsom salts, iodine, Band-Aids and so on — her concern and patience had been unlimited, and her cures often worked. Much later we learned that many people had thought that she possessed magic powers for healing. Thus she had secured for me the goodwill of the local human population.

Above me, Fifi stirred, cradling little Flossi more comfortably as she suckled. Then her eyes closed again. The infant nursed for a few more minutes, then the nipple slipped from her mouth as she too slept. I continued to daydream, re-living in my mind some of the more memorable events of the past.

I remembered the day when David Greybeard had first visited my camp by the lakeshore. He had come to feed on the ripe fruits of an oil-nut palm that grew there, spied some bananas on the table outside my tent, and taken them off to eat in the bush. Once he had discovered bananas he had returned for more and gradually other chimpanzees had followed him to my camp.

One of the females who became a regular visitor in 1963 was Fifi's mother, old Flo of the ragged ears and bulbous nose. What an exciting day when, after five years of maternal preoccupation with her infant daughter, Flo had become sexually attractive again. Flaunting her shell-pink sexual swelling she had attracted a whole retinue of suitors. Many of them had never been to camp, but they had followed Flo there, sexual passions overriding natural caution. And, once they had discovered bananas, they had joined the rapidly growing group of regular camp visitors. And so I had become more and more familiar with the whole host of unforgettable chimpanzee characters who are described in my first book, *In the Shadow of Man*.

Fifi, lying so peacefully above me now, was one of the few survivors of those early days. She had been an infant when first I knew her in 1961. She had weathered the terrible polio epidemic that had swept through the population — chimpanzee and human alike — in 1966. Ten of the chimpanzees of the study group had died or vanished.

Another five had been crippled, including her eldest brother, Faben, who had lost the use of one arm.

At the time of that epidemic the Gombe Stream Research Centre was in its infancy. The first two research assistants were helping to collect and type out notes on chimp behavior. Some twenty-five chimpanzees were regularly visiting camp by then, and so there had been more than enough work for all of us. After watching the chimps all day we had often transcribed notes from our tape recorders until late at night.

My mother Vanne had made two other visits to Gombe during the sixties. One of those had been when the National Geographic Society sent Hugo van Lawick to film the study — which, by then, they were financing. Louis Leakey had wangled Vanne's fare and expenses, insisting that it would not be right for me to be alone in the bush with a young man. How different the moral standards of a quarter of a century ago! Hugo and I had married anyway, and Vanne's third visit, in 1967, had been to share with me, for a couple of months, the task of raising my son, Grub (his real name is Hugo Eric Louis) in the bush.

There was a slight movement from Fifi's nest and I saw that she had turned and was looking down at me. What was she thinking? How much of the past did she remember? Did she ever think of her old mother, Flo? Had she followed the desperate struggle of her brother, Figan, to rise to the top-ranking, alpha position? Had she even been aware of the grim years when the males of her community, often led by Figan, had waged a sort of primitive war against their neighbours, assaulting them, one after the other, with shocking brutality? Had she known about the gruesome cannibalistic attacks made by Passion and her adult daughter Pom on newborn infants of the community?

Again my attention was jerked back to the present, this time by the sound of a chimpanzee crying. I smiled. That would be Fanni. She had reached the adventurous age when a young female often moves away from her mother to travel with the adults. Then, suddenly, she wants mother desperately, leaves the group, and sets off to search for her. The crying grew louder and soon Fanni came into sight. Fifi paid no attention, but Flossi jumped out of the nest and scrambled down to embrace her elder sister. And Fanni, finding Fifi where she had left her, stopped her childish crying.

Clearly Fifi had been waiting for Fanni — now she climbed down and set off, and the children, followed after, playing as they went. The family moved rapidly down the steep slope to the south. As I scrambled after them, every branch seemed to catch in my hair or my shirt. Frantically I crawled and wriggled through a terrible tangle of undergrowth. Ahead of me the chimpanzees, fluid black shadows, moved effortlessly. The distance between us increased. The vines curled around the buckles of my shoes and the strap of my camera, the thorns caught in the flesh of my arms, my eyes smarted till the tears flowed as I yanked my hair from the snags that reached out from all around. After ten minutes I was drenched in sweat, my shirt was torn, my knees bruised from crawling on the stony ground — and the chimps had vanished. I kept quite still, trying to listen above the pounding of my heart, peering in all directions through the thicket around me. But I heard nothing.

For the next thirty-five minutes I wandered along the rocky bed of the Kasakela Stream, pausing to listen, to scan the branches above me. I passed below a troop of red colobus monkeys, leaping through the tree tops, uttering their strange, high-pitched, twittering calls. I encountered some baboons of D troop, including old Fred with his one blind eye and the double kink in his tail. And then, as I was wondering where to go next, I heard the scream of a young chimp further up the valley. Ten minutes later I had joined Gremlin with little Galahad, Gigi and two of Gombe's youngest and most recent orphans, Mel and Darbee, both of whom had lost their mothers when they were only just over three years old. Gigi, as she so often does these days, was 'auntying' them both. They were all feeding in a tall tree above the almost dry stream and I stretched out on the rocks to watch them. During my scramble after Fifi the sun had vanished, and now, as I looked up through the canopy, I could see the sky, grey and heavy with rain. With a growing darkness came the stillness, the hush, that so often precedes hard rain. Only the rumbling of the thunder, moving ever closer, broke this stillness; the thunder and the rustling movements of the chimpanzees.

When the rain began Galahad, who had been dangling and patting at his toes near his mother, quickly climbed to the shelter of her arms. And the two orphans hurried to sit, close together, near Gigi. But Gimble started leaping about in the tree tops, swinging vigorously from one branch to the next, climbing up then jumping down to catch himself on a bough below. As the rain got heavier, as more and more drops found their way through the dense canopy, so his leaps became wilder and ever more daring, his swaying of the branches more vigorous. This behaviour would, when he was older, express itself in the magnificent rain display, or rain dance, of the adult male.

Suddenly, just after three o'clock, heralded by a blinding flash of lightning and a thunderclap that shook the mountains and growled on and on, bouncing from peak to peak, the grey-black clouds let loose such torrential rain that sky and earth seemed joined by moving water. Gimble stopped playing then, and he, like the others, sat hunched and still, close to the trunk of the tree. I pressed myself against a palm, sheltering as best I could under its overhanging fronds. As the rain poured down endlessly I got colder and colder. Soon, turned in upon myself, I lost all track of time. I was no longer recording — there was nothing to record except silent, patient and uncomplaining endurance.

It must have taken about an hour before the rain began to ease off as the heart of the storm swept away to the south. At 4.30 the chimps climbed down, and moved off through the soaked, dripping vegetation. I followed, walking awkwardly, my wet clothes hindering movement. We travelled along the stream bed then up the other side of the valley, heading south. Presently we arrived on a grassy ridge overlooking the lake. A pale, watery sun had appeared and its light caught the raindrops so that the world seemed hung with diamonds, sparkling on every leaf, every blade of grass. I crouched low to avoid destroying a jewelled spider's web that stretched, exquisite and fragile, across the trail.

The chimpanzees climbed into a low tree to feed on fresh young leaves. I moved to a place where I could stand and watch as they enjoyed their last meal of the day. The scene was breathtaking in its beauty. The leaves were brilliant, a pale, vivid green in the soft sunlight;

the wet trunk and branches were like ebony; the black coats of the chimps were shot with flashes of coppery-brown. And behind this vivid tableau was the dramatic backcloth of the indigo-black sky where the lightning still flickered and flashed, and the distant thunder rumbled.

There are many windows through which we can look out into the world, searching for meaning. There are those opened up by science, their panes polished by a succession of brilliant, penetrating minds. Through these we can see ever further, ever more clearly, into areas that once lay beyond human knowledge. Gazing through such a window I have, over the years, learned much about chimpanzee behaviour and their place in the nature of things. And this, in turn, has helped us to understand a little better some aspects of human behaviour, our own place in nature.

But there are other windows; windows that have been unshuttered by the logic of philosophers; windows through which the mystics seek their visions of the truth; windows from which the leaders of the great religions have peered as they searched for purpose not only in the wondrous beauty of the world, but also in its darkness and ugliness. Most of us, when we ponder on the mystery of our existence, peer through but one of these windows onto the world. And even that one is often misted over by the breath of our finite humanity. We clear a tiny peephole and stare through. No wonder we are confused by the tiny fraction of a whole that we see. It is, after all, like trying to comprehend the panorama of the desert or the sea through a rolled-up newspaper.

As I stood quietly in the pale sunshine, so much a part of the rain-washed forests and the creatures that lived there, I saw for a brief moment through another window and with other vision. It is an experience that comes, unbidden, to some of us who spend time alone in nature. The air was filled with a feathered symphony, the evensong of birds. I heard new frequencies in their music and, too, in the singing of insect voices, notes so high and sweet that I was amazed. I was intensely aware of the shape, the colour, of individual leaves, the varied patterns of the veins that made each one unique. Scents were clear, easily identifiable — fermenting, over-ripe fruit; water-logged earth; cold, wet bark; the damp odour of chimpanzee hair and, yes, my own too. And the aromatic scent of young, crushed leaves was almost over-powering. I sensed the presence of a bushbuck, then saw him, quietly browsing upwind, his spiralled horns dark with rain. And I was utterly filled with that peace 'which passeth all understanding'.

Then came far-off pant-hoots from a group of chimpanzees to the north. The trance-like mood was shattered. Gigi and Gremlin replied, uttering their distinctive pant-hoots. Mel, Darbee and little Galahad joined in the chorus.

I stayed with the chimps until they nested — early, after the rain. And when they had settled down, Galahad cosy beside his mother, Mel and Darbi each in their own small nests close to the big one of auntie Gigi, I left them and walked back along the forest trail to the lakeshore. I passed the D troop baboons again. They were gathered around their sleeping trees, squabbling, playing, grooming together, in the soft light of evening. My walking feet crunched the shingle of the beach, and the sun was a huge red orb above the lake. As it lit the clouds for yet another magnificent display, the water became golden, shot with gleaming ripples of violet and red below the flaming sky.

Later, as I crouched over my little wood fire outside the house, where I had cooked, then eaten, beans and tomatoes and an egg, I was still lost in the wonder of my experience that afternoon. It was, I thought, as though I had looked onto the world through such a window as a chimpanzee might know. I dreamed, by the flickering flames. If only we could, however briefly, see the world through the eyes of a chimpanzee, what a lot we should learn.

A last cup of coffee and then I would go inside, light the hurricane lamp, and write out my notes of the day, the wonderful day. For, since we cannot know with the mind of a chimpanzee we must proceed laboriously, meticulously, as I have for thirty years. We must continue to collect anecdotes and, slowly, compile life histories. We must continue, over the years, to observe, record and interpret. We have, already, learned much. Gradually, as knowledge accumulates, as more and more people work together and pool their information, we are raising the blind of the window through which, one day, we shall be able to see even more clearly into the mind of the chimpanzee.

2

The Mind of
the Chimpanzee

<<<<<<<<<<<<<<

FTEN I HAVE GAZED into a chimpanzee's eyes and wondered what was going on behind them. I used to look into
Flo's, she so old, so wise. What did she remember of her
young days? David Greybeard had the most beautiful eyes of them
all, large and lustrous, set wide apart. They somehow expressed his
whole personality, his serene self-assurance, his inherent dignity —
and, from time to time, his utter determination to get his way. For a
long time I never liked to look a chimpanzee straight in the eye — I
assumed that, as is the case with most primates, this would be inter-
preted as a threat or at least as a breach of good manners. Not so. As
long as one looks with gentleness, without arrogance, a chimpanzee
will understand, and may even return the look. And then — or such
is my fantasy — it is as though the eyes are windows into the mind.
Only the glass is opaque so that the mystery can never be fully revealed.

I shall never forget my meeting with Lucy, an eight-year-old home-
raised chimpanzee. She came and sat beside me on the sofa and, with
her face very close to mine, searched in my eyes — for what? Perhaps
she was looking for signs of mistrust, dislike, or fear, since many people
must have been somewhat disconcerted when, for the first time, they
came face to face with a grown chimpanzee. Whatever Lucy read in
my eyes clearly satisfied her for she suddenly put one arm round my
neck and gave me a generous and very chimp-like kiss, her mouth
wide open and laid over mine. I was accepted.

For a long time after that encounter I was profoundly disturbed. I had been at Gombe for about fifteen years then and I was quite familiar with chimpanzees in the wild. But Lucy, having grown up as a human child, was like a changeling, her essential chimpanzeeness overlaid by the various human behaviours she had acquired over the years. No longer purely chimp yet eons away from humanity, she was man-made, some other kind of being. I watched, amazed, as she opened the refrigerator and various cupboards, found bottles and a glass, then poured herself a gin and tonic. She took the drink to the TV, turned the set on, flipped from one channel to another then, as though in disgust, turned it off again. She selected a glossy magazine from the table and, still carrying her drink, settled in a comfortable chair. Occasionally, as she leafed through the magazine she identified something she saw, using the signs of ASL, the American Sign Language used by the deaf. I, of course, did not understand, but my hostess, Jane Temerlin (who was also Lucy's 'mother'), translated: 'That dog,' Lucy commented, pausing at a photo of a small white poodle. She turned the page. 'Blue,' she declared, pointing then signing as she gazed at a picture of a lady advertising some kind of soap powder and wearing a brilliant blue dress. And finally, after some vague hand movements — perhaps signed mutterings — 'This Lucy's, this mine,' as she closed the magazine and laid it on her lap. She had just been taught, Jane told me, the use of the possessive pronouns during the thrice weekly ASL lessons she was receiving at the time.

The book written by Lucy's human 'father', Maury Temerlin, was entitled *Lucy, Growing Up Human*. And in fact, the chimpanzee is more like us than is any other living creature. There is close resemblance in the physiology of our two species and genetically, in the structure of the DNA, chimpanzees and humans differ by only just over one per cent. This is why medical research uses chimpanzees as experimental animals when they need substitutes for humans in the testing of some drug or vaccine. Chimpanzees can be infected with just about all known human infectious diseases including those, such as hepatitis B and AIDS, to which other non-human animals (except gorillas, orangutans and gibbons) are immune. There are equally striking similarities between humans and chimpanzees in the anatomy and wiring of the brain and nervous system, and — although many scientists have been reluctant to admit to this — in social behaviour, intellectual ability, and the emotions. The notion of an evolutionary continuity in physical structure from pre-human ape to modern man has long been morally acceptable to most scientists. That the same might hold good for mind was generally considered an absurd hypothesis — particularly by those who used, and often misused, animals in their laboratories. It is, after all, convenient to believe that the creature you are using, while it may react in disturbingly human-like ways, is, in fact, merely a mindless and, above all, unfeeling, 'dumb' animal.

When I began my study at Gombe in 1960 it was not permissible — at least not in ethological circles — to talk about an animal's mind. Only humans had minds. Nor was it quite proper to talk about animal personality. Of course everyone knew that they *did* have their own unique characters — everyone who had ever owned a dog or other pet was aware of that. But ethologists, striving to make theirs a 'hard' science, shied away from the task of trying to explain such things objectively. One respected ethologist, while acknowledging that there was 'variability between individual animals', wrote that it was best that this fact be 'swept under the carpet'. At that time ethological carpets fairly bulged with all that was hidden beneath them.

How naive I was. As I had not had an undergraduate science education I didn't realize that animals were not supposed to have personalities, or to think, or to feel emotions or pain. I had no idea that it would have been more appropriate to assign each of the chimpanzees a number rather than a name when I got to know him or her. I didn't realize that it was not scientific to discuss behaviour in terms of motivation or purpose. And no one had told me that terms such as *childhood* and *adolescence* were uniquely human phases of the life cycle, culturally determined, not to be used when referring to young chimpanzees. Not knowing, I freely made use of all those forbidden terms and concepts in my initial attempt to describe, to the best of my ability, the amazing things I had observed at Gombe.

I shall never forget the response of a group of ethologists to some remarks I made at an erudite seminar. I described how Figan, as an adolescent, had learned to stay behind in camp after senior males had left, so that we could give him a few bananas for himself. On the first occasion he had, upon seeing the fruits, uttered loud, delighted food calls: whereupon a couple of the older males had charged back, chased after Figan, and taken his bananas. And then, coming to the point of the story, I explained how, on the next occasion, Figan had actually suppressed his calls. We could hear little sounds, in his throat, but so quiet that none of the others could have heard them. Other young chimps, to whom we tried to smuggle fruit without the knowledge of their elders, never learned such self-control. With shrieks of glee they would fall to, only to be robbed of their booty when the big males charged back. I had expected my audience to be as fascinated and impressed as I was. I had hoped for an exchange of views about the chimpanzee's undoubted intelligence. Instead there was a chill silence, after which the chairman hastily changed the subject. Needless to say, after being thus snubbed, I was very reluctant to contribute any comments, at any scientific gathering, for a very long time. Looking back, I suspect that everyone was interested, but it was, of course, not permissible to present a mere 'anecdote' as evidence for anything.

The editorial comments on the first paper I wrote for publication demanded that every *he* or *she* be replaced with *it*, and every *who* be replaced with *which*. Incensed, I, in my turn, crossed out the *its* and *whichs* and scrawled back the original pronouns. As I had no desire to carve a niche for myself in the world of science, but simply wanted to go on living among and learning about chimpanzees, the possible reaction of the editor of the learned journal did not trouble me. In fact I won that round: the paper when finally published did confer upon the chimpanzees the dignity of their appropriate genders and properly upgraded them from the status of mere 'things' to essential Being-ness.

However, despite my somewhat truculent attitude, I did want to learn, and I was sensible of my incredible good fortune in being admitted to Cambridge. I wanted to get my PhD, if only for the sake of Louis Leakey and the other people who had written letters in support of my admission. And how lucky I was to have, as my supervisor, Robert Hinde. Not only because I thereby benefitted from his brilliant mind and clear thinking, but also because I doubt that I could have found a teacher more suited to my particular needs and personality. Gradually he was able to cloak me with at least some of the trappings of a scientist. Thus although I continued to hold to most of my convictions — that animals had personalities; that they could feel happy or sad or fearful; that they could feel pain; that they could strive towards planned goals and achieve greater success if they were highly motivated — I soon realized that these personal convictions were, indeed, difficult to prove. It was best to be circumspect — at least until I had gained some credentials and credibility. And Robert gave me wonderful advice on how best to tie up some of my more rebellious ideas with scientific ribbon. 'You can't *know* that Fifi was jealous,' he admonished on one occasion. We argued a little. And then: 'Why don't you just say *If Fifi were a human child we would say she was jealous.*' I did.

It is not easy to study emotions even when the subjects are human. I know how I feel if I am sad or happy or angry, and if a friend tells me that he is feeling sad, happy or angry, I assume that his feelings are similar to mine. But of course I cannot know. As we try to come to grips with the emotions of beings progressively more different from ourselves the task, obviously, becomes increasingly difficult. If we ascribe human emotions to non-human animals we are accused of being anthropomorphic — a cardinal sin in ethology. But is it so terrible? If we test the effect of drugs on chimpanzees because they are biologically so similar to ourselves, if we accept that there are dramatic similarities in chimpanzee and human brain and nervous system, is it not logical to assume that there will be similarities also in at least the more basic feelings, emotions, moods of the two species?

In fact, all those who have worked long and closely with chimpanzees have no hesitation in asserting that chimps experience emotions similar to those which in ourselves we label pleasure, joy, sorrow, anger, boredom and so on. Some of the emotional states of the chimpanzee are so obviously similar to ours that even an inexperienced observer can understand what is going on. An infant who hurls himself screaming to the ground, face contorted, hitting out with his arms at any nearby object, banging his head, is clearly having a tantrum. Another youngster, who gambols around his mother, turning somersaults, pirouetting and, every so often, rushing up to her and tumbling into her lap, patting her or pulling her hand towards him in a request for tickling, is obviously filled with *joie de vivre*. There are few observers who would not unhesitatingly ascribe his behaviour to a happy, carefree state of well-being. And one cannot watch chimpanzee infants for long without realizing that they have the same emotional need for affection and reassurance as human children. An adult male, reclining in the shade after a good meal, reaching benignly to play with an infant or idly groom an adult female, is clearly in a good mood. When he sits with bristling hair, glaring at his subordinates and threatening them, with irritated gestures, if they come too close, he is clearly feeling cross and grumpy. We make these judgements because the similarity of so much of a chimpanzee's behaviour to our own permits us to empathize.

It is hard to empathize with emotions we have not experienced. I can imagine, to some extent, the pleasure of a female chimpanzee during the act of procreation. The feelings of her male partner are beyond my knowledge — as are those of the human male in the same context. I have spent countless hours watching mother chimpanzees interacting with their infants. But not until I had an infant of my own did I begin to understand the basic, powerful instinct of mother-love. If someone accidentally did something to frighten Grub, or threaten his well-being in any way, I felt a surge of quite irrational anger. How much more easily could I then understand the feelings of the chimpanzee mother who furiously waves her arm and barks in threat at an individual who approaches her infant too closely, or at a playmate who inadvertently hurts her child. And it was not until I knew the numbing grief that gripped me after the death of my second husband that I could even begin to appreciate the despair and sense of loss that can cause young chimps to pine away and die when they lose their mothers.

Empathy and intuition can be of tremendous value as we attempt to understand certain complex behavioural interactions, provided that the behaviour, as it occurs, is recorded precisely and objectively. Fortunately I have seldom found it difficult to record facts in an orderly manner even during times of powerful emotional involvement. And 'knowing' intuitively how a chimpanzee is feeling — after an attack, for example — may help one to understand what happens next. We should not be afraid at least to try to make use of our close evolutionary relationship with the chimpanzees in our attempts to interpret complex behaviour.

Today, as in Darwin's time, it is once again fashionable to speak of and study the animal mind. This change came about gradually, and was, at least in part, due to the information collected during careful studies of animal societies in the field. As these observations became widely known, it was impossible to brush aside the complexities of social behaviour that were revealed in species after species. The untidy clutter under the ethological carpets was brought out and examined, piece by piece. Gradually it was realized that parsimonious explanations of apparently intelligent behaviours were often misleading. This led to a succession of experiments that, taken together, clearly prove that many intellectual abilities that had been thought unique to humans were actually present, though in a less highly developed form, in other, non-human beings. Particularly, of course, in the non-human primates and especially in chimpanzees.

When first I began to read about human evolution, I learned that one of the hallmarks of our own species was that we, and only we, were capable of making tools. *Man the Toolmaker* was an oft-cited definition — and this despite the careful and exhaustive research of Wolfgang Kohler and Robert Yerkes on the tool-using and tool-making abilities of chimpanzees. Those studies, carried out independently in the early twenties, were received with scepticism. Yet both Kohler and Yerkes were respected scientists, and both had a profound understanding of chimpanzee behaviour. Indeed, Kohler's descriptions of the personalities and behaviour of the various individuals in his colony, published in his book *The Mentality of Apes*, remain some of the most vivid and colourful ever written. And his experiments, showing how chimpanzees could stack boxes, then climb the unstable constructions to reach fruit suspended from the ceiling, or join two short sticks to make a pole long enough to rake in fruit otherwise out of reach, have become classic, appearing in almost all textbooks dealing with intelligent behaviour in non-human animals.

By the time systematic observations of tool-using came from Gombe those pioneering studies had been largely forgotten. Moreover, it was one thing to know that humanized chimpanzees in the lab could use implements: it was quite another to find that this was a naturally occurring skill in the wild. I well remember writing to Louis about my first observations, describing how David Greybeard not only used bits of straw to fish for termites but actually stripped leaves from a stem and thus *made* a tool. And I remember too receiving the now oft-quoted telegram he sent in response to my letter: 'Now we must re-define *tool*, redefine *Man*, or accept chimpanzees as humans.'

There were, initially, a few scientists who attempted to write off the termiting observations, even suggesting that I had taught the chimps! By and large, though, people were fascinated by the information and by the subsequent observations of the other contexts in which the Gombe chimpanzees used objects as tools. And there were only a few anthropologists who objected when I suggested that the chimpanzees probably passed their tool-using traditions from one generation to the next, through observations, imitation and practice, so that each population might be expected to have its own unique tool-using culture. Which, incidentally, turns out to be quite true. And when I described how one chimpanzee, Mike, spontaneously solved a new problem by using a tool (he broke off a stick to knock a banana to the ground when he was too nervous to actually take it from my hand) I don't believe there were any raised eyebrows in the scientific community. Certainly I was not attacked viciously, as were Kohler and Yerkes, for suggesting that humans were not the only beings capable of reasoning and insight.

The mid-sixties saw the start of a project that, along with other similar research, was to teach us a great deal about the chimpanzee mind. This was Project Washoe, conceived by Trixie and Allen Gardner. They purchased an infant chimpanzee and began to teach her the signs of ASL, the American Sign Language used by the deaf. Twenty years earlier another husband and wife team, Richard and Cathy Hayes, had tried, with an almost total lack of success, to teach a young chimp, Vikki, to talk. The Hayes's undertaking taught us a lot about the chimpanzee mind, but Vikki, although she did well in IQ tests, and was clearly an intelligent youngster, could not learn human speech. The Gardners, however, achieved spectacular success with their pupil, Washoe. Not only did she learn signs easily, but she quickly began to string them together in meaningful ways. It was clear that each sign evoked, in her mind, a mental image of the object it represented. If, for example, she was asked, in sign language, to fetch an apple, she would go and locate an apple that was out of sight in another room.

Other chimps entered the project, some starting their lives in deaf signing families before joining Washoe. And finally Washoe adopted an infant, Loulis. He came from a lab where no thought of teaching signs had ever penetrated. When he was with Washoe he was given no lessons in language acquisition — not by humans, anyway. Yet by the time he was eight years old he had made fifty-eight signs in their correct contexts. How did he learn them? Mostly, it seems, by imitating the behaviour of Washoe and the other three signing chimps, Dar, Moja and Tatu. Sometimes, though, he received tuition from Washoe herself. One day, for example, she began to swagger about bipedally, hair bristling, signing *food! food! food!* in great excitement. She had seen a human approaching with a bar of chocolate. Loulis, only eighteen months old, watched passively. Suddenly Washoe stopped her swaggering, went over to him, took his hand, and moulded the sign for *food* (fingers pointing towards mouth). Another time, in a similar context, she made the sign for *chewing gum* — but with *her* hand on *his* body. On a third occasion Washoe, apropos of nothing, picked up a small chair, took it over to Loulis, set it down in front of him, and very distinctly made the *chair* sign three times, watching him closely as she did so. The two food signs became incorporated into Loulis's vocabulary but the sign for chair did not. Obviously the priorities of a young chimp are similar to those of a human child!

When news of Washoe's accomplishments first hit the scientific community it immediately provoked a storm of bitter protest. It implied that chimpanzees were capable of mastering a human language, and this, in turn, indicated mental powers of generalization, abstraction and concept-formation as well as an ability to understand and use abstract symbols. And these intellectual skills were surely the prerogatives of *Homo sapiens*. Although there were many who were fascinated and excited by the Gardners' findings, there were many more who denounced the whole project, holding that the data was suspect, the methodology sloppy, and the conclusions not only misleading, but quite preposterous. The controversy inspired all sorts of other language projects. And, whether the investigators were sceptical to start with and hoped to disprove the Gardners' work, or whether they were attempting to demonstrate the same thing in a new way, their research provided additional information about the chimpanzee's mind.

And so, with new incentive, psychologists began to test the mental abilities of chimpanzees in a variety of different ways; again and again the results confirmed that their minds are uncannily like our own. It had long been held that only humans were capable of what is called 'cross-modal transfer of information' — in other words, if you shut your eyes and someone allows you to feel a strangely shaped potato, you will subsequently be able to pick it out from other differently shaped potatoes simply by looking at them. And vice versa. It turned out that chimpanzees can 'know' with their eyes what they 'feel' with their fingers in just the same way. In fact, we now know that some other non-human primates can do the same thing. I expect all kinds of creatures have the same ability.

Then it was proved, experimentally and beyond doubt, that chimpanzees could recognize themselves in mirrors — that they had, therefore, some kind of self-concept. In fact, Washoe, some years previously, had already demonstrated the ability when she spontaneously identified herself in the mirror, staring at her image and making her name sign. But that observation was merely anecdotal. The proof came when chimpanzees who had been allowed to play with mirrors were, while anaesthetized, dabbed with spots of odourless paint in places, such as the ears or the top of the head, that they could see only in the mirror. When they woke they were not only fascinated by their spotted images, but immediately investigated, with their fingers, the dabs of paint.

The fact that chimpanzees have excellent memories surprised no one. Everyone, after all, has been brought up to believe that 'an elephant never forgets' so why should a chimpanzee be any different?

The fact that Washoe spontaneously gave the name-sign of Beatrice Gardner, her surrogate mother, when she saw her after a separation of eleven years was no greater an accomplishment than the amazing memory shown by dogs who recognize their owners after separations of almost as long — and the chimpanzee has a much longer life span than a dog. Chimpanzees can plan ahead, too, at least as regards the immediate future. This, in fact, is well illustrated at Gombe, during the termiting season: often an individual prepares a tool for use on a termite mound that is several hundred yards away and absolutely out of sight.

This is not the place to describe in detail the other cognitive abilities that have been studied in laboratory chimpanzees. Among other accomplishments chimpanzees possess pre-mathematical skills: they can, for example, readily differentiate between *more* and *less*. They can classify things into specific categories according to a given criterion — thus they have no difficulty in separating a pile of food into *fruits* and *vegetables* on one occasion, and, on another, dividing the same pile of food into *large* versus *small* items, even though this requires putting some vegetables with some fruits. Chimpanzees who have been taught a language can combine signs creatively in order to describe objects for which they have no symbol. Washoe, for example, puzzled her caretakers by asking, repeatedly, for a *rock berry*. Eventually it transpired that she was referring to Brazil nuts which she had encountered for the first time a while before. Another language-trained chimp described a cucumber as a *green banana*, and another referred to an Alka-Seltzer as a *listen drink*. They can even invent signs. Lucy, as she got older, had to be put on a leash for her outings. One day, eager to set off but having no sign for *leash*, she signalled her wishes by holding a crooked index finger to the ring on her collar. This sign became part of her vocabulary. Some chimpanzees love to draw, and especially to paint. Those who have learned sign language sometimes spontaneously label their works, 'This [is] apple' — or bird, or sweetcorn, or whatever. The fact that the paintings often look, to our eyes, remarkably unlike the objects depicted by the artists either means that the chimpanzees are poor draughtsmen or that we have much to learn regarding ape-style representational art!

People sometimes ask why chimpanzees have evolved such complex intellectual powers when their lives in the wild are so simple. The answer is, of course, that their lives in the wild are not so simple! They use — and need — all their mental skills during normal day-to-day life in their complex society. They are always having to make choices — where to go, or with whom to travel. They need highly developed social skills — particularly those males who are ambitious to attain high positions in the dominance hierarchy. Low-ranking chimpanzees must learn deception — to conceal their intentions or to do things in secret — if they are to get their way in the presence of their superiors. Indeed, the study of chimpanzees in the wild suggests that their intellectual abilities evolved, over the millennia, to help them cope with daily life. And now, the solid core of data concerning chimpanzee intellect collected so carefully in the lab setting provides a background against which to evaluate the many examples of intelligent, rational behaviour that we see in the wild.

It is easier to study intellectual prowess in the lab where, through carefully devised tests and judicious use of rewards, the chimpanzees can be encouraged to exert themselves, to stretch their minds to the limit. It is more meaningful to study the subject in the wild, but much harder. It is more meaningful because we can better understand the environmental pressures that led to the evolution of intellectual skills in chimpanzee societies. It is harder because, in the wild, almost all behaviours are confounded by countless variables; years of observing, recording and analysing take the place of contrived testing; sample size can often be counted on the fingers of one hand; the only experiments are nature's own, and only time — eventually — may replicate them.

In the wild a single observation may prove of utmost significance, providing a clue to some hitherto puzzling aspect of behaviour, a key to the understanding of, for example, a changed relationship. Obviously it is crucial to see as many incidents of this sort as possible. During the early years of my study at Gombe it became apparent that one person alone could never learn more than a fraction of what was going on in a chimpanzee community at any given time. And so, from 1964 onwards, I gradually built up a research team to help in the gathering of information about the behaviour of our closest living relatives.

3

The Research Centre

THE GOMBE STREAM RESEARCH CENTRE grew from small beginnings to become one of the most dynamic field stations for the study of animal behaviour in the world. The first two research assistants joined me in 1964. It was not long before we found that there was more work than we three could manage, even though Hugo, my husband, was there to help as well. And so we sought additional funds to employ additional students. Almost all of them fell under the spell of Gombe and repaid our faith in them by helping us to collect more and ever more information about the lives of the chimpanzees.

By 1972 there were sometimes as many as twenty students, for by then we were studying not only chimpanzees, but baboons as well. There were graduate students from a variety of disciplines, mainly anthropology, ethology and psychology, from universities in the United States and Europe. And there were undergraduates too, from the interdisciplinary human biology programme at Stanford University and from the zoology department of the University of Dar es Salaam. The students slept in separate miniports — little aluminium huts hidden away among the trees near camp — but everyone gathered together in the mess for meals. This was a functional cement and stone building down on the beach, built by my old friend George Dove, at whose camp on the Serengeti Hugo and I had stayed when Grub was a baby. George had built offices, too, and a kitchen with a wood stove. And he had installed a generator so that we could have some electricity: this meant that we could work more easily at night and also enabled us to operate a deep freeze that made catering less of a nightmare. George even built a little stone house for us to use as a dark room.

Life at the research centre was busy. In addition to the main business of observing the animals and collecting data, there were weekly seminars at which we discussed research findings and planned ever better ways of collating the information from the various studies. There was a spirit of cooperation among the students, a willingness to share data, that was, I think, quite unusual. It had not been easy to foster this generous attitude — initially many of the graduate students were, understandably, reluctant to contribute any of their precious data to a central information pool. But clearly this had to be done if we were to come to grips with the extraordinarily complex social organization of the chimpanzees and document as fully as possible their life histories. I was helped not only by many of the students themselves, but also by Dave Hamburg, head of the department of psychiatry at Stanford University. It was he who had brought in the human biology students. And although these young people seldom stayed more than six months at Gombe, they had been so well prepared before they arrived in Africa that their contributions were very valuable.

Most important of all for the long-term future of the research at Gombe, though we did not know it then, was the training of the Tanzanian field staff. From 1968, when one of the students fell over a cliff while following chimps, and tragically lost her life, it had become the custom for each student to be accompanied in the bush by a local Tanzanian. Then, if there was an accident, one of the two could go for help. Gradually these men had acquired knowledge that made their assistance invaluable: they knew all the chimpanzees by name and could identify them for newcomers, and they were experts at finding their way around the rugged terrain. By 1972 they had begun to collect data themselves — for example, marking the travel route of a target chimpanzee on a map, noting who he or she associated with during the day, and identifying the various food plants that were eaten. The graduate students relied quite heavily on this pool of data, and so they worked hard to ensure that the field assistants were well trained. From time to time I held seminars in Kiswahili, the language used across

East Africa, during which we discussed various aspects of chimpanzee and baboon behaviour, and I talked about other non-human primates in different parts of the world. And so the field staff gradually became better informed, more interested and more enthusiastic.

I felt immensely proud to have been responsible for bringing this group of people together, and the quality and quantity of information that was being gathered was extraordinary. Yet there were times when I thought back to my early days at Gombe with real nostalgia — the very early days, when my only companions were my mother, Dominic the cook, and Hassan who drove the little motor boat into Kigoma for supplies. I had worked incredibly hard, forcing myself to climb to the Peak at dawn and remaining out until the mountains were already shadowed by the coming night. There were no weekends, no holidays. But I was young and physically fit and I gloried in it. And I was on my own. I could travel through the forests knowing that the only beings I would meet all day would be chimpanzees, or baboons, or some of the other wild creatures that inhabit the lush valleys or the more open mountain slopes. But change had been inevitable: there was no way in the world that one person, no matter how dedicated, could have made a really comprehensive study of the Gombe chimpanzees. Hence the research centre, the growing number of people moving about in the forests, the decreasing likelihood of spending hours at a time in absolute solitude.

In truth, by 1972 I was spending only very short periods with the chimpanzees, despite the fact that, apart from the three months a year that I spent teaching a course in the human biology programme at Stanford, I was living permanently at Gombe. This was because, after spending the previous few years watching chimpanzee mothers raising their infants, I was trying my hand at bringing up a child myself. It had become quite clear to me that a close, affectionate bond with the mother was important for the future well-being of a young chimpanzee. I suspected that the same was true for humans, and the work of men such as Rene Spitz and John Bowlby confirmed this. I was determined to give my own son the best start I could. And so, while the students spent most of their time in the field, I spent most of my time with Grub. (Although his real name is Hugo, he is still known as Grub to his family and closest friends even now.) I usually worked at administration and writing in the morning and did things with Grub in the afternoons.

Of course I kept abreast of all that was going on in the chimpanzee community. The conversation each evening, in the mess, was very rarely about anything other than chimpanzees and baboons. I was able to follow, albeit vicariously, the dominance rivalry between Humphrey, Figan and Evered. I received daily bulletins on the adolescent exploits of Flint and Goblin, Pom and Gilka, and on Gigi's sexual adventures. Moreover, I almost always saw at least one or two chimps during my daily visits to camp.

Occasionally Grub and I had chimpanzee visitors at our house on the beach. Once Melissa and her family wandered along the veranda and peered through the weldmesh windows of the living room, just after someone had brought Grub two pet rabbits. There are no rabbits at Gombe, and the chimpanzees were clearly fascinated. Goblin, filled with the intense curiosity of adolescence, continued to hang onto the window, staring and staring, long after his mother and little sister had lost interest and left. Incidentally, they were terrific pets, those rabbits, quite house-trained, very affectionate and extremely entertaining. And they taught me a lot — I had no idea until then, for example, that rabbits enjoyed meat. And I was even more startled to watch them hunting and eating spiders!

Chimpanzees have been known to seize and eat human infants, and so that Grub would have maximum safety Hugo and I had built our house on the beach as the chimpanzees seldom went there. The baboons, however, were often on the lakeshore, and our house was in the heart of the range of Beach troop. As a result, I spent more time watching baboons than I ever had before. This was not only a good learning experience in itself, but it gave me a new perspective on chimpanzee behaviour, pinpointing ways in which it differs from that of monkeys, such as baboons. Chimpanzees are clearly more 'intellectual' than baboons — as demonstrated by their use of objects as tools, for example. But baboons are very much more adaptive than chimps. There are baboons all over Africa, from north to south, east to west, whereas the chimpanzees, with their cautious and conservative natures and their much slower reproductive rate, are found only in the equatorial forest belt and surrounding areas.

From the very first, the baboons at Gombe, bold and opportunistic, were quick to try any imported human foods they could get their hands on — and almost without exception, they found them highly desirable commodities. There has been a constant battle of wits between the humans at Gombe on the one hand and baboons on the other — a battle won, only too often, by the baboons. In vain did we make rules: no food to be eaten outside; no food remains to be thrown out except in covered rubbish pits; food that has to be carried from one place to another must be covered; house doors must be closed at all times. Everyone tried to obey the rules but there were always times when someone forgot, or was in a hurry, or thought, 'Well, there aren't any baboons around now.' And those are the moments baboons wait for.

The baboon Crease was an inveterate thief. He used to sit patiently for hours, concealed in some thickly foliaged tree behind one of our houses, far from the rest of his troop. If we forgot to latch the door, even for a few moments, he would seize the opportunity to make a quick raid. Many a loaf of bread, handful of eggs, pineapple or paw-paw did he snatch from the shelves before we imposed heavy fines on careless behaviour that led to such depredations. Once he stole a two-pound tin of margarine, newly opened, and sat, consuming the contents slowly and with apparent relish, for the next two hours.

One day Grub, highly excited, told me an epic Crease story. It began when a water taxi (as we call the little boats that carry passengers up and down the lake) broke down near the research centre. The boat was pulled up to the edge of the beach, the outboard engine was taken off for repair, and the passengers got out to stretch their legs. Somehow Crease got wind of the fact that there was a load of cassava (manioc) flour on the empty boat. Without hesitation the old reprobate jumped aboard. But even as he ripped open one of the sacks, and began stuffing the food into his mouth, the boat started to drift out into the lake. Then, suddenly noticing that the shore was receding, Crease panicked. As he leaped from one side of the boat to the other he kept bumping into the ripped sack so that clouds of white dust rose from it, making him sneeze. Finally one of the students took pity on him and, weak with laughter, pulled the boat back to shore. Crease disembarked with undignified haste, frosted like a Christmas decoration.

In fact baboons, unlike chimpanzees, can swim. Sometimes, when the water is calm, the young baboons go into the lake for fun, even diving down and swimming underwater. During aggressive incidents a baboon will sometimes escape from its persecutors by running out into the lake and waiting there until things have quietened down.

Lake Tanganyika is said to be the largest body of uncontaminated water anywhere: it is the longest lake in the world and the second deepest. Great storms sometimes sweep its length, stirring the surface into huge waves. Almost every year a few fishermen are blown miles out towards Zaire, some never to return. And there are other dangers, too, lurking in the crystal depths of the lake. The crocodiles have gone now, but there are water cobras living among the great rocks that march out into the water at the headland of each bay. There is no anti-venom that will save you if you are bitten by one of these long, sleek brown snakes, with black bands around its neck. That was why I always worried about Grub when he was swimming in the lake. Yet in most ways Gombe was a wonderful environment for bringing up a child.

Grub spent much of his early childhood pottering about on the shores of the lake, and it was probably there, surrounded by traditional fishermen, that he acquired his passion for fishing. As a small boy he showed unbelievable patience when it came to untangling some hope-lessly snarled fishing net. Whereas I would have given up after the first few minutes, he would persist for a whole morning, and sometimes into the afternoon, until at last the net was neatly laid out on the veranda, complete with floats, ready to set before nightfall. And then, after the excitement of examining the catch the next morning, the whole laborious process had to be gone through again.

When Grub was five years old, he began a school correspondence course under the direction of a series of tutors — young people filling in a year between school and university, glad of the opportunity to see Gombe and the chimps in return for their services. But there was still much opportunity for fishing and swimming in the lake. It was at this same time that Maulidi Yango came into Grub's life. Maulidi, who was employed to help with the cutting of trails through the forest, has a splendid physique and is as strong as an ox. Newcomers to

Gombe would sometimes be startled to see what appeared to be an entire tree moving ahead of them along some trail: and then, somewhere under the tree, they would see Maulidi! Easy-going, with a great sense of humour, Maulidi became Grub's childhood hero. Indeed, Grub maintains that Maulidi had more influence in shaping his character than anyone else outside the family. It was a commonplace sight at Gombe to see Maulidi stretched out on the sand while Grub swam, Maulidi paddling a canoe while Grub fished — or Maulidi eating his midday meal and enjoying his midday siesta while Grub waited. They have remained firm friends.

One morning Grub came to tell me that Flo and Flint were near the mess. By this time Flo was a very old lady indeed. Her teeth were worn down to the gums and she had trouble finding enough soft foods to eat. We gave her extra banana rations in camp and when she came near the house I always gave her eggs. But even so she gradually became more and more frail. Still, from time to time, she showed flashes of the indomitable spirit that, undoubtedly, had enabled her to live to such a ripe old age.

So it was that morning. I found her sitting on the ground, hunched and looking cold and miserable, for it had rained a short while before — one of those short, heavy deluges that sometimes catch one unawares in the middle of the dry season. Close by, Flint was teasing Crease. The old baboon was minding his own business, but Flint kept shaking rain-laden branches above his head, showering him with drops. In the end Crease, who had been bowing his head as though trying to ignore Flint, lost his temper and leapt up at his tormentor, threatening him. Flint screamed, and at once Flo sprang into action. Sticking her few remaining moth-eaten hairs on end she charged at Crease, uttering fierce waa-barks of threat. And Crease fled!

A few weeks later, Crease tried to take one of the eggs I had just given to Flo. She bristled up at once, stood upright, and ran at the baboon, flailing her arms and actually hitting him. And Crease withdrew and sat watching from a respectful distance as the ancient female slowly savoured the eggs, one at a time, chewing them with leaves.

Sometimes I followed Flo and Flint when they wandered past the house. From time to time, Flint still tried to ride on his old mother and she would have carried him, I believe, had she been physically strong enough. As it was, she collapsed under his weight and so he had to walk. Even without him on her back Flo had to sit and rest frequently during travel, and Flint often became impatient, moving on and then whimpering when she did not immediately follow. Sometimes he went up to her and, with a sullen pout on his face, pushed her vigorously, trying to force her to move on. When she insisted on resting he gave her no peace but constantly pestered her for grooming, pulling her hands towards him, crying petulantly if she refused. Once he even pulled her out of a low day nest, so that she tumbled ignominiously onto the ground. Often I felt like slapping him. Yet it was clear that Flo would have been very lonely without him. She moved so slowly that even her daughter Fifi seldom travelled with her, and by then Flo had become almost as dependent on Flint as he was on her. I remember once, when they came to a fork in the trail, Flo went one way, Flint the other. I followed Flo. After a few minutes she stopped, looked back, and gave a few low, sad whimpers. She waited a while, hoping I suppose that Flint would change his mind. When he did not appear, she turned and went after her son.

It was a bright, clear morning when I received news of her death. Her body had been found, lying face down in the Kakombe Stream. Although I had long known that the end was close, this did nothing to mitigate the grief that filled me as I stood looking down at Flo's remains. I had known her for eleven years and I had loved her.

I watched over her body that night, to keep marauding bushpigs from violating it. Flint was still nearby, and his grief might have been the worse had he found his mother's body torn and partly eaten. As I kept my vigil in the bright moonlight, I thought about Flo's life. For nigh on fifty years she must have roamed the Gombe hills. And even if I had not arrived to record her history, to invade the privacy of that rugged terrain, Flo's life would have been, in and of itself, significant and worthwhile, filled with purpose, vigour, and love of life. And how much I had learned from her during her long acquaintance. For she taught me to honour the role of the mother in society, and to appreciate not only the immeasurable importance to a child of good mothering but also the utter joy and contentment which that relationship can bring to the mother.

4

Mothers and Daughters

<<<<<<<<<<<<

'MANNERS MAKYTH MAN,' wrote the poet William of Wykeham. Ah — but what makyth the manners? We might, perhaps, venture 'Mother makyth manners' — along, of course, with a dash of early experience and more than a little spicing of genetic inheritance. The relative roles of 'nature' versus 'nurture' caused much bitter argument in scientific circles in recent years. But the flames of the controversy have now died down, and it is generally accepted that, even in the lower animals, adult behaviour is acquired through a mix of genetic make-up and experience gained as the individual goes through life. The more complex an animal's brain the greater the role that learning is likely to play in shaping its behaviour, and the more variation we shall find between one individual and another. Information acquired and lessons learned during infancy and childhood, when behaviour is at its most flexible, are likely to have particular significance.

For chimpanzees, whose brains are more like those of humans than are those of any other living animal, the nature of early experience may have a profound effect on adult behaviour. Particularly important, I believe, is the disposition of the child's mother, his or her position in the family, and, if there are elder siblings, their sex and personalities. A secure childhood is likely to lead to self-reliance and independence in adulthood. A disturbed early life may leave permanent scars. In the wild almost all mothers look after their infants relatively efficiently. But even so there are clear-cut differences in the child-raising techniques of different individuals. It would be hard to find two females whose mothers had treated them more differently during their early years, than Flo's daughter Fifi and Passion's daughter Pom. In fact, Flo and Passion are at opposite ends of a scale: most mothers fall somewhere between these two extremes.

Fifi had a carefree — a wonderful — childhood. Old Flo was a highly competent mother, affectionate, tolerant, playful and protective. Figan was an integral part of the family when Fifi was growing up, joining her games when Flo was not in the mood and often supporting his young sister in her childhood squabbles. Faben, Flo's oldest son, was often around too. Flo, who held top rank among the females when I first knew her, was a sociable female. She spent a good deal of time with other members of her community, and she had a relaxed and friendly relationship with most of the adult males. In this social environment Fifi became a self-confident and assertive child.

Pom's childhood, in comparison with Fifi's, was bleak. Passion's personality was as different from Flo's as chalk from cheese. Even when I first knew her in the early sixties she was a loner. She had no close female companions, and on those occasions when she was in a group with adult males her relationship with them was typically uneasy and tense. She was a cold mother, intolerant and brusque, and she seldom played with her infant, particularly during the first two years. And Pom, being the first surviving child, had no sibling to play with during the long hours when she and her mother were on their own. She had a difficult time during her early months, and she became an anxious and clinging child, always fearful that her mother would go off and leave her behind.

Thus it is not really surprising that Pom and Fifi reacted differently to the various challenges that a young female must face as she grows up in the wild.

All chimpanzee infants become upset and depressed during the difficult time of weaning when the mother prevents her child, with increasing frequency and determination, both from suckling and from riding on her back. This usually takes place during the fourth year. Fifi became noticeably less cheerful and less playful for a few months and she spent more and more time sitting in close contact with her mother, looking hunched and sad. But she got over her depression quickly, and by the time her infant brother Flint was born, was back to her old self — outgoing, confident and assertive.

Pom's depression, however, seemed to go on for ever. Interestingly, sometime during her daughter's third year, Passion's attitude towards her had softened: she had become more patient and more playful. And Pom, presumably as a direct result of this, had gradually become less anxious. But these signs of improved psychological well-being disappeared during the trauma of weaning. It was clearly a far more disturbing experience for Pom than it had been for Fifi, despite the fact that Passion, to my surprise, was remarkably tolerant. She almost always responded to Pom's frequent requests for grooming and even allowed her to ride on her back with a minimum of protest. For weeks after we were sure that her milk had dried up, she let Pom sit close, a nipple in her mouth, her eyes often closed, for as long as twenty minutes at a time. But nothing seemed to help. Pom's inability to cope with weaning was almost certainly due to the harsh treatment she had received as an infant. So often her only succour had been her mother's milk and now, when this was suddenly denied, her early sense of insecurity returned. It was not until a few weeks before Passion gave birth to her next infant that Pom finally quit trying to suckle from her mother.

For all young chimps the birth of a new baby in the family signals the end of an era, a major step towards independence — although it will be another three to six years before they begin to leave their mothers and move out into the adult world. Fifi was about five and a half years old when Flint was born. Now that Flo had a tiny infant to care for she could not give her undivided attention to Fifi. But far from being upset, Fifi was utterly fascinated and delighted by the new baby, and spent hours, during his first two years, playing with him, grooming him, and carrying him during family travel. She jealously chased off other youngsters when they wanted to play with him, at least when he was small, and helped Flo by retrieving him from potentially dangerous situations.

Pom, like Fifi, was initially curious and fascinated when infant Prof was born. But soon, after the novelty of her little brother had worn off, she reverted to the depressed state in which she had been before his birth. And she remained lethargic and listless for most of Prof's first year of life, seldom showing much interest in him. Even when, at five months old, he began to toddle about — a stage that Fifi had found irresistible — Pom remained unresponsive to Prof. She seldom carried him, and when they played, which was not often, the game was usually initiated by Prof. Gradually, however, Pom got over her depression, and her brother then became more appealing. She began to carry him and play with him more often. She became very protective, too. Once, for example, as Pom led her family through the forest, she noticed a large snake coiled up beside the trail. Uttering a small warning 'huu', she swung up into a tree. Three-year-old Prof, tottering along behind his sister, seemed not to see the snake. If he did, he had no thought of possible danger. Nor, apparently, did he understand Pom's soft warning. Passion, bringing up the rear, was far behind. Suddenly, when Prof was within a few yards of the snake, Pom, every hair on end with fright, rushed down, gathered up her little brother, and climbed with him to safety.

The next major upheaval in the life of a young female chimpanzee is when, at about ten years of age, she becomes for the first time sexually attractive to the big males. Fifi was enchanted by this new experience. Sometimes, when a male was, quite obviously, uninterested in what she had to offer, she would recline close by and, ever hopeful, stare at him. Or rather, stare at a certain portion of his anatomy that was, so far as she was concerned, disappointingly flabby. Once she went so far as to tweak the limp appendage — with highly satisfactory results! It soon became clear that the males regarded Fifi as a most desirable sexual partner. She did not have quite the sex appeal that Flo had once radiated — but in those days she was, after all, younger and less experienced.

When Pom, in her turn, first became sexually attractive to the adult males she, like Fifi, clearly found the new experience pleasurable and hastened to any male who showed signs of interest. But whereas Fifi had been calm and relaxed when she complied with the sexual demands of the males, Pom crouched before them, tense and nervous, and the moment intercourse was over she leapt away, often screaming. She developed strange, neurotic behaviours. Often, for example, as she went up to a male to greet him, she would utter loud and frenzied pant-barks of submission and, crouching in front of him, dab a hand out towards his face, then leap away. The males were irritated by this and sometimes threatened or even attacked her. And so, in a vicious circle, her nervousness and tension increased. It was scarcely surprising that Pom was far less popular as a sexual partner than Fifi had been at the same age.

Adolescent female chimpanzees, like their human counterparts, typically go through an infertile phase between menarche and the first conception. For both Fifi and Pom this period lasted for about two years — two years during which, for about ten days each month, they came into oestrus and were sexually attractive and highly receptive to the adult males. These months were clearly beneficial to Fifi. Although Flo sometimes accompanied her daughter when she went in search of male company, she was old, and Fifi often went without her. And so she learned how to get on in adult society without having to rely on support from her high-ranking mother. As she matured socially and became more self-reliant, she filled out and became stronger too — she would be the more able to cope when she eventually became a mother herself.

Nevertheless, while Fifi became increasingly independent and worldly-wise, she always rejoined her mother after each period of dalliance with the males. And so she was still very much a part of the family when, in 1968, Flo gave birth to her last baby. Sadly, little Flame only lived for six months, but during that time Fifi, whenever she had the opportunity — when she was not sexually preoccupied with the males — delighted to carry, groom and gently play with the tiny infant, thus gaining additional experience in maternal skills.

Towards the end of her two-year period of infertility, Fifi was frequently taken off by one or other of her male suitors to the outskirts of the community range. There the couple would remain — if the male could pull it off — isolated from other males, for the duration of Fifi's swelling. It is during such consortships that males have a good chance of siring a child. In fact, though, it is fairly certain that Fifi's first infant was not fathered by a male of her own community, but by one of the Kalande males in the south — for Fifi made a number of visits to their territory, obeying the peculiar urge to wander, to meet and mate with stranger males, that we have observed in most females during late adolescence. And it seems that she conceived during one of those excursions. Once pregnant, Fifi returned to her own home range. Her relationship with Flo and seven-year-old Flint became even closer now that her sexual urge was, for a while, quiescent.

Pom's adolescence was more turbulent. By then the bond between herself and her mother was very close indeed — even closer, in some ways, than that between Fifi and Flo. Passion would always back up her daughter during squabbles with other community females and Pom had become assertive and aggressive in her dealings with them. When Passion was not close by, the others would often retaliate, picking a fight with Pom. But if Passion was close enough to hear her daughter's screaming she would race to her defence and mother and daughter together would then punish the female concerned. And Pom typically tried to help and support her mother in the same way.

One such incident is clear in my mind. I had followed Pom all morning and was watching as she and another female, Nope, fished for termites. Presently we heard pant-hoots, then some screams, about half a mile to the west, further down the valley. Both females stared towards the sounds but while Nope at once returned to her feeding, Pom continued to gaze westward. After a few moments there was another outburst of calling. Nope paid no attention, but Pom gave a little grin of fear, reached to touch Nope, and continued to look towards the distant group. A minute later came the frenzied screaming of a chimpanzee being attacked. Instantly, with a squeak of fear, Pom was off, racing towards the sounds. There was, fortunately for me, a rough trail and I was not left too far behind. We ran for about five hundred yards and then, as I burst through a tangle of vines, I saw that Pom had joined her mother and was grooming her. Both Passion and Prof, who was up in a tree above, were bleeding from fresh wounds, received, without doubt, during the attacks we had just heard. An adult male charged towards us, hit out at Passion and her daughter, then charged away, leaving the family alone.

Even during the periods when Pom was pink and went off in search of sexual gratification, Passion often went with her. And if Pom did travel with the males on her own, she usually returned quite soon to the reassuring company of her mother and little Prof. Not until her sixth pinkness was Pom observed to sleep with a group of males far away from her family.

Unlike Fifi, Pom was seldom taken on consortships — and at least part of the reason lay in her unusually close relationship with Passion. I well remember one hot September afternoon in 1976, when at midday I found Pom accompanied, as usual, by her mother and brother. With them was Satan — and he was trying most desperately to lead Pom away to the north. Pom, however, had no desire to go with him. Again and again, hair bristling, eyes glaring, Satan swayed vegetation at the young female, then moved off in his chosen direction, looking back to see if she was following. Again and again Pom ignored these summonses. Several times, exasperated, Satan swaggered around Pom, threatening her. And when this happened Pom, screaming loudly, rushed to Passion for comfort. Then Passion, tough old bird that she was, glared at the big male and uttered a string of angry — and surely abusive — waa-barks. Once Satan attacked Pom, and immediately Passion, with furious barks, hurled herself at her daughter's assailant, hitting him with her fists. Satan was probably as surprised as I was! He left the daughter and turned on the mother — but he only attacked her mildly. Passion and Pom then groomed each other for a long time while Satan sat, glowering, nearby. After this he made only two more swaggering attempts to impose his will — and then, nearly four hours after I had first encountered them, he gave up and went off on his own. Pom had been well chaperoned!

The birth of a first baby, is, for the mother, an event of epic significance. And in Fifi's case the birth was of great significance for me also. Indeed, during the eight months of Fifi's pregnancy I was almost (but not quite!) as impatient as I had been during my own pregnancy four years earlier. Would she, as I predicted, be the same kind of mother as Flo? We first saw her baby in May 1971 when he was about two days old. Remembering the wild sexual adventures of his mother's adolescence, we named him — naturally enough — Freud! Just as ex-

pected, Fifi was from the start a relaxed and competent mother. Like
Flo before her, she was tolerant, affectionate and playful. And she too
showed some of the behaviour that had been unique to her mother.

One day, when Freud was just a few months old, a student called
to me: 'Isn't that what Flo used to do?' And there was Fifi dangling
Freud from one foot while she tickled him just as Flo had played with
Flint! Until then, no other mother had been observed to play in that
special way. Fifi had tried when, as a child, she played with little Flint,
but in those days her legs had been too short. Now she imitated Flo
to perfection.

Fifi continued to spend most of her time with her own mother during
Freud's first year of life, but disappointingly, Flo showed little interest
in her grandson. Sometimes she peered at him and, as he grew older,
she tolerated him when he occasionally held onto her hair. But by then
Flo was very old indeed; she had barely enough energy to get her frail
body through each day and there was none left over for luxuries such
as playing with her daughter's infant. Freud was only fifteen months
old when his grandmother died.

And what of Pom and her first baby? She was almost exactly thirteen
years old when Pan was born. I had expected her to treat him much
as she herself had been treated as an infant but in this case (fortunately
for Pan) my predictions were largely wrong. Pom was most definitely
a more attentive and tolerant mother than Passion had been. Indeed,
when first I watched her with her baby, carefully supporting him during travel whenever he lost his grip, it seemed that she had the makings
of a really considerate mother. But there was something lacking: Pom
did not develop anything like the degree of maternal proficiency and
concern shown by Fifi.

Indeed, in some ways Pom's behaviour did reflect the manner in
which she herself had been handled as an infant. She found it difficult
to cradle Pan comfortably when he was small — or else she simply
couldn't be bothered. Often, as she sat in a tree, the infant would slip
down off her lap and hang on frantically with wildly kicking legs as
he tried to pull himself back up again. Only when he whimpered did
Pom look down and, appearing slightly surprised, gather him back
onto her thighs. But she seldom made any attempt to make a better

lap and often, after a few minutes, he slipped down again and the sequence was repeated. Pom, like Passion, tended to move off without first gathering her infant; but unlike Passion, Pom almost always hurried back at his first whimper of distress. It seemed that she always expected Pan would be able to follow, but was instantly concerned when she found that he could not. Pom, like Passion, was not a playful mother, but Pan did not suffer as a result since Pom continued to spend most of her time with Passion and her new infant, Pax. And Pax, just a year older than Pan, was the perfect playmate.

Pom, for all that she was a far better mother than I had expected, lost this first child. I was there to witness the horrifying accident that led to his death. It was one of those violently blustery mornings in August when the wind roars down the valley in great gusts, tossing the tree tops and sweeping on to wreak havoc over the lake. For about half an hour I had been lying on my back watching Pom and Pan as they fed on oil nuts forty-five feet above me. Pan was almost three years old, able to poke the occasional fruit from its horny case though preferring to beg for a half-chewed one from his mother. For a while he clung tightly to Pom's hair, made nervous, as most chimps are, by the violent wind. But then he got bold and ventured further afield despite the gale. Suddenly a really fierce gust lashed savagely at the fronds and Pan, like a stuffed toy, was swept from the tree. He seemed almost to float through the air, his arms and legs spread-eagled, as though he was lying flat out on some buoyant but invisible air mattress. As he hit the ground, rock hard after the fierce suns of summer, there was a sickening thud. A moment later came two strangled, heart-wrenching exhalations, then silence.

I was shaking when I moved toward his body. He lay as he had fallen, on his back. His eyes were closed. I looked up at Pom, left alone so suddenly in the tree. She was staring down at the ground. Very slowly, as though afraid, she climbed down and approached her infant. Cautiously she reached out, and gathered up the tiny form. To my utter astonishment he gripped her hair and clung, unaided, as she moved away. I had been certain he was already dead.

For the next two hours Pom rested and groomed her infant. No mother could have shown more concern, more solicitude. Pan suckled for a long time, then leaned against Pom, with his eyes closed. When he did move, it was very slowly and, not surprisingly, he seemed quite dazed. I assumed he was, at the very least, suffering from concussion. Presently Pom gathered up her battered child and carried him up into a tall tree to feed.

Unfortunately this happened the day I was due to leave Gombe. The boat was waiting and I could not follow the tragedy through to the end. Three days later, when Pom was next seen, Pan was dead. Presumably he sustained internal injuries or a fractured skull — or both. By a strange coincidence, three weeks later, in Dar es Salaam, a little boy, the seven-year-old son of my neighbour's cook, fell from a coconut palm and landed, like Pan, on his back. He was rushed to the hospital where they found extensive internal damage, including a ruptured liver. They patched him up as best they could, but he too died a short while afterwards.

It would be unfair to blame Pom entirely for the accident, to accuse her of negligence. It could have happened to any infant. Yet I cannot imagine Fifi losing a child in this way. For Fifi, like Flo before her, like all really attentive chimpanzee mothers, is alert to potential danger. Often she 'rescues' her infant before the child itself has shown any sign of distress or fear. After Pan's death, I began to watch carefully whenever Fifi, with one of her infants, fed up in a palm tree during a strong wind. Always the infant stayed close to her. Although I could not determine whether that was due to Fifi's concern or the apprehension of the child, in some ways it comes to the same thing: if the infant is extra cautious it is probably at least in part because its movements have been firmly restricted in similar circumstances in the past.

Pom, after the tragic death of little Pan, became sick, lethargic and so emaciated we thought she might not recover. Her relationship with her mother now became, if anything, even closer, and they were seldom apart. I remember one day when they did accidentally become separated. Pom searched for Passion for almost an hour, frequently whimpering softly to herself, and from time to time climbing tall trees and gazing out from these vantage points in all directions. To some extent she may have been helped by occasional whiffs of Passion's characteristic odour for, as she travelled, she repeatedly bent and sniffed the trail or picked up leaves and smelled them carefully before dropping them. When eventually mother and daughter were reunited, Pom rushed up to Passion with small squeaks of excitement and pleasure, and the two groomed for over an hour.

As we shall see, the life histories of Fifi and Pom have continued along very different lines. Pom, after her mother's death, became increasingly solitary and eventually left the community for good. Fifi, by contrast, has become one of the most high-ranking and respected females in her group, maintaining close friendly relations with the adult males and many females too. She has also become the most reproductively successful Kasakela female to date. Whether Flo's main contribution to Fifi was genetic or through child-raising skills or through the equal mix of the two, the recipe worked. And her two eldest sons, who also received fifty per cent of their genes from their mother and who were probably brought up in much the same way, thrived on Flo's recipe too. Particularly the younger of the two, Figan, who became, for a while, the most powerful alpha male in Gombe's recorded history.

5

Figan's Rise

‹‹‹‹‹‹‹‹‹‹‹‹

FROM THE OUTSET it was obvious that Figan was endowed with exceptional intelligence: I gave many examples in my earlier book, *In the Shadow of Man*. Equally clear was his determination to attain an ever higher position in male society. He developed an impressive charging display. This display serves to make a chimpanzee look bigger and more dangerous than he may actually be — his hair stands on end; he leaps up to shake the vegetation; he drags huge branches noisily along the ground, then hurls them ahead of him; he picks up and throws rocks with such vigour that they fly unpredictably ahead, behind or to the side; he stamps and slaps loudly upon the ground or some tree trunk; his lips are tightly compressed, pulling his face into a ferocious scowl. And the wilder and more impressive his display, and the more carefully it is planned and executed, the better his chance of intimidating his rivals without recourse to actual physical combat — during which he himself, as well as his opponent, might be injured. The smaller the individual the more it behooves him to work on his display.

Even as an adolescent Figan was quick to notice and try to take advantage of any sign of weakness (such as sickness or injury) in one of the adult males. Then, while the higher-ranked individual was at a disadvantage, Figan hurled his challenge — his impressive charging display — again and again. Often he was ignored, even threatened. But sometimes his audacity paid off and the older male, at least until he had recovered, would hasten out of his way. Even a temporary victory of this sort served to increase Figan's self-confidence.

When Mike deposed Goliath and rose to the top-ranking position of the community Figan was eleven years old and, clearly, fascinated by the imaginative strategy of the new alpha. For Mike, by incorporating empty four-gallon tin cans into his charging displays, hitting and kicking them ahead of him as he ran towards his rivals, succeeded in intimidating them all — including individuals much larger than himself. All the chimps were impressed by these unique, noisy and often terrifying performances. But Figan was the only one whom we saw, on two different occasions, 'practising' with cans that had been abandoned by Mike. Characteristically — for he was a past master at keeping out of trouble — he did this only when out of sight of the older males who would have been intolerant of such behaviour in a mere adolescent. He would undoubtedly have become as adroit as Mike had we not removed all cans from circulation.

Figan's strong motivation to better his social position, along with his intelligence, earmarked him as a future alpha. The only serious drawback seemed to be his very highly strung nature. During intense social excitement, for example, he sometimes began to scream uncontrollably and often rushed over to a nearby individual, touching or embracing him, or her, for reassurance. Sometimes he even clutched his own scrotum. Nevertheless, as I was finishing *In the Shadow of Man* I wrote: 'I suspect that Figan will eventually become the top-ranking male.'

The story that lies behind Figan's long struggle to the top is a fascinating one. It revolves around the complex and changing relationships between himself and three other males — his brother, Faben; his childhood playmate, Evered; and, oldest of the four, the powerful and unusually aggressive Humphrey.

When Faben was striken by polio and lost the use of one arm, Figan managed to dominate his older brother. For the next three years the two young males interacted very little. Indeed, had they not been equally drawn to spend time with their mother they would probably have drifted apart, for Faben, at that time, was friendly with Humphrey, and Figan was clearly ill at ease in the presence of the much larger and stronger male.

Then, when Figan was sixteen years old, the nature of his relationship with Faben changed again. The brothers became increasingly friendly and, for the first time, we observed them joining forces against one of Figan's rivals, his childhood playmate Evered. Together the brothers defeated him with ease, and wounded him quite badly into the bargain.

For some time prior to that attack, relations between Figan and Evered had been strained and tense. When they met they had often performed vigorous charging displays as each tried to intimidate the other. Evered, by virtue of his seniority, had usually come off best, but after his defeat by the brothers he began to greet Figan with nervous panting grunts whenever they met. At least, he did for a few months. Youth, however, can be resilient, and Evered, like Figan, was also highly motivated to climb the social ladder. Gradually Evered's confidence returned — partly, no doubt, because Figan was by no means always with his brother: Faben was still friendly with Humphrey, and Figan, wisely, steered clear of the powerful male. Moreover, even when the brothers were together, Faben did not *always* help Figan: sometimes he just sat and watched.

By that time, although Mike was still top-ranking, he was showing signs of age. His teeth were worn, the canines broken. His hair, dull and brown, was beginning to thin. It is not surprising that Figan, perceptive and astute as ever, was the first to challenge the authority of the failing alpha. Initially he merely ignored Mike's charging displays: he sat facing the other way! This clearly had an unnerving effect on Mike, who sometimes displayed again and again in Figan's vicinity as though desperately trying to provoke some sign of respect. But Figan was not impressed and, as the weeks went by, he himself displayed ever more frequently when he was near Mike. And soon Evered also began to question Mike's position.

Both these young males, however, continued to show extreme deference to Humphrey. And Humphrey himself, through sheer force of custom (since he could have defeated Mike hands down in actual combat) was still highly respectful of the old alpha. Thus in 1969 I wrote: 'Soon, then, we may have a situation where no single male is dominant in all situations. Certainly something is going to happen very soon.'

Something did, on a dull, grey day early in January, 1970. Mike was sitting in camp by himself, peacefully eating a few bananas, when suddenly Humphrey, closely followed by Faben, charged up the slope and attacked him — just like that. For no obvious reason, with no apparent provocation. Mike, screaming, sought refuge up a tree. Humphrey followed, pulled him to the ground, and hit and stamped on him again. Faben, for good measure, joined the fray, and pounded on Mike a couple of times. Humphrey, seeming almost shocked by what he had done, was already leaving, and Faben followed him. The two aggressors vanished, leaving Mike utterly shattered, giving soft calls of fear and distress.

It had all happened so suddenly, was over so quickly. Yet it was truly an historic event, for it marked the end of an era, Mike's six-year reign as alpha. Almost overnight he became one of the lowest-ranking males of his community: even some of the adolescents began to challenge him, and Mike seldom tried to stand up for himself.

A week after his defeat, I followed the fallen monarch when he left camp. He moved slowly, pausing often to pick and munch on various leaves and fruits along the way. Later, in the heat of midday, he bent a few saplings onto the ground and settled down on this little bed to rest. I leaned against the trunk of a gnarled old fig nearby. It was quiet and very peaceful. Mike lay, his eyes open, staring into space. As I watched him I wondered what was going on in his mind. Was he regretting his lost power? Is it only we humans, with our constant pre-occupation with self-image, who know the crippling sense of humiliation? Mike turned his head and looked at me, looked directly into my eyes. His gaze seemed untroubled, serene. Perhaps, I thought, he was glad to relax and let go the reins of power. After all, it is hard work for a top-ranking chimpanzee to maintain his position even when he is strong and young. And Mike was so old, so tired. Presently he closed his eyes and slept. Later, when he awoke, he wandered off into the forest, a solitary figure, very small under the huge trees.

Humphrey automatically succeeded Mike as alpha. But although it was a decisive victory he had won, it was hardly glorious. He was strong and in his prime. He weighed at least twenty pounds more than the aging Mike. No grim determination to succeed, no hard-won series of battles against a powerful adversary, lay behind this rise to top rank. And, despite his large build and fiery temperament, Humphrey never became a truly impressive alpha: he was little more than a blustering bully, lacking the drive, intelligence and courage that had been so impressive in both Mike and his predecessor Goliath.

Indeed, but for a lucky break — the departure of Hugh and Charlie, the two males whom he feared the most — Humphrey would never have made it to the top at all. This had happened a few months before Humphrey defeated Mike, at a time when the community that I had been observing for ten years began to divide. Some of them were spending ever more time in the far south of the range which, until then, all members of the community had shared. The leaders of the move to the south were Hugh and Charlie. Almost certainly brothers, the two had a close, supportive relationship and almost always travelled about together. They made a formidable team and it was hardly surprising that Humphrey, who had no close friend and only the occasional support of one-armed Faben, was fearful of them. When Hugh and Charlie, along with the other 'southern' males, made one of their occasional excursions back to the north, Humphrey was usually able to avoid them. Gradually these expeditions became less and less frequent and eventually stopped altogether.

Everything seemed to be going Humphrey's way. Not only was he rid of his main rivals but, as a result of the community division, there were now only eight adult males over whom he had to maintain control: Mike, and Goliath before him, had had to exercise authority over up to fourteen. Yet despite this auspicious start, Humphrey only held his top-ranking position for one and a half years. He was usurped by Figan.

Even during the early months of his reign Humphrey seemed to sense, in Figan, potential danger: he displayed, bristling and magnificent, much more often in Figan's presence than at other times. Probably such performances served to boost his own self-confidence, as well as to impress Figan. Figan, for his part, initially continued to keep out of Humphrey's way as much as possible and was, at least outwardly, highly respectful of the new alpha. Meanwhile he was still preoccupied with his long struggle to dominate Evered. Indeed, looking back on the events of the stormy period it seems probable that Figan, all along, realized that Evered, rather than Humphrey, was his most formidable rival.

Soon after the change in alpha males a serious fight took place between Evered and Figan. As the two males skirmished high in a tree, Evered was joined by one of the senior males, and Figan, outmatched, fell some thirty feet to the ground. Evered, victorious, displayed magnificently through the branches while Figan sat screaming below. He was badly hurt, having sprained his wrist or, perhaps, broken some small bone in his hand, and he was very lame for the next three weeks.

This happened just two months before Flo's death. She looked incredibly ancient; her body was shrunken, her eyes, for the most part, dull and blank, her movements slow. Yet when she heard the frenzied screaming of her son, at least a quarter of a mile away, she leapt to her feet and, with all her remaining hairs standing on end, raced towards the sounds — so fast that her human follower was left far behind. When she arrived on the scene there seemed little she could do, this frail old lady, to help Figan against his powerful aggressors. But her very presence calmed him. His frantic screaming gave way to soft whimpers as he limped towards his mother. And when she began to groom him he quietened altogether, relaxing under the reassuring touch of her fingers just as he had throughout his infancy and childhood. When Flo moved off, Figan followed, holding his bad hand off the ground. Not until his injury had healed did he leave her and move back into adult male society with all its tensions and dangers, its excitement and exhilaration.

The next recorded drama was a fight between Figan and Humphrey. It was not very dramatic, and neither male was hurt, but it marked, for the alpha male, the beginning of the end. When it was over each of the combatants repeatedly ran to touch or embrace one of the other males present. They were not only seeking reassurance, but also trying to enlist allies. In this only Figan was successful: he persuaded one or two of the others to join him and, together, they charged at Humphrey who fled and, for several days, is thought to have wandered by himself. His period of greatest control had ended — but Figan's had not yet begun.

The more we learn about the struggle for power among chimpanzees, the more we realize the tremendous importance of coalitions. An adult male trying to make it to the top has a much better chance of success if he has an ally — a friend who will consistently come to his assistance in times of need and, even more important from a psychological point of view, who will not side with a rival against him.

A temporary alliance now sprang up between Humphrey and Evered. They sought each other's company and became frequent grooming partners. When they were together, each giving the other moral support, they could afford to ignore Figan's tempestuous displays. Indeed, they jointly defeated him in a fight a few months later. But this did not change things much — Humphrey, for the most part, avoided Figan, while the tension and hostility between Figan and Evered seemed, if anything, to increase. The charging displays that each performed in the vicinity of the other, when they met, became ever more vigorous. Once they charged back and forth, first one and then the other, for the best part of an hour. Figan, hair bristling, ran toward Evered, hurled a large rock, and displayed past him, scattering other members of the group. Then he sat, out of breath. A few moments later Evered started up. He leapt to shake and sway vegetation near his rival, dragged a branch past him, then in his turn sat panting from his exertions. Five minutes later, Figan began another performance. And so it went on. They created much excitement and nervous tension among their spectators before they finally gave up, probably from exhaustion. So far as we could tell, the score at the end of that round was even.

Figan, despite his intelligence and his desire for high rank, might never have attained the coveted alpha position but for a sudden change of heart in Faben. Up until that time, although Faben has almost never joined sides *against* his younger brother, he had by no means always supported him either. But all at once, towards the end of 1972, the relationship between the two became even closer: if Figan challenged another male, Faben, if present, would join in, displaying in unison with his brother. If Figan needed help, Faben was prepared to give it. He became, it seemed, utterly committed to supporting Figan in his quest for power.

Why did Faben show this sudden change of heart? Was it perhaps, at least in part, a consequence of Flo's death? The closer bond between the brothers was not apparent immediately following her passing, but then neither Faben (nor Figan for that matter) saw her dead body so there was no way of knowing, at the time, that Flo had vanished for ever. Then, as weeks went by with no sign of her, may not Faben have begun to feel a creeping sense of loss, an empty place in his heart, full-grown male though he was? A certain loneliness which he tried to assuage by spending more time with his brother?

Certainly Faben as well as Figan had, as an adult, found comfort in his mother's familiar, unthreatening presence. Once, when he hurt his foot, Faben (like Figan when he sprained his wrist) had travelled with Flo until he was well again. There was also the time when Faben returned, after a long sojourn in the north, with the hand of his par-alysed arm badly infected. He was, quite clearly, in considerable pain. He moved very slowly, walking upright and cradling the swollen fingers with his good hand. For several days he remained close to camp, constantly scanning the slopes of the valley, as though looking for someone. We shall never know whether, as I suspect, he was seeking comfort from his mother, for Flo, by one of those ironic twists of fate, had died the day before his return.

Whatever the reasons behind Faben's decision to whole-heartedly support his younger brother, by April 1973 the two were all but in-separable. It was the strength of this alliance that not only brought about Humphrey's final downfall, but enabled Figan, at long last, to vanquish Evered, too. He accomplished these victories during three major conflicts.

The first of these took place at the end of April. Figan and Faben jointly attacked Evered, who took refuge up a tree, whimpering and screaming. The brothers continued to charge about below for over half an hour until, during a lull, their victim finally managed to escape.

Four days later came the second. This time Figan tackled Hum-phrey — a far more dangerous opponent than Evered when it came to actual fighting, since Humphrey weighed at least fifteen pounds more than either Figan or Evered. It happened in the evening. All four main characters were present — indeed they had been together all day in a large mixed group, feasting on the lush crops that abound at the end of the long wet season. There had been the usual kinds of excitement — charging displays and squabbles. Nothing out of the ordinary. As the sun sank low toward the lake in the west, Figan was feeding by himself, some distance from the others. The sound of snapping branches and rustling leaves indicated that the chimpanzees were beginning to make their nests for the night. It was a peaceful time, a time for gentle relaxation after the long day, before stretching out with a full belly.

Figan stopped feeding. For a few moments he sat motionless in his tree and then, quite calmly, he climbed down. But by the time he reached the others his hair had begun to bristle and, as he climbed their tree, moving ever faster, he swelled until he seemed twice his normal size. Suddenly he was off, displaying wildly through the branches, swaying them violently, leaping and swinging from one side of the tree to the other. There was instant pandemonium as chimpanzees screamed and fled his approach, many of them leaping from their nests. Figan briefly chased an old male, swatted him in passing and then, having worked himself into a frenzy, leapt down onto Humphrey where he sat in his nest. The two males, locked in combat, fell at least thirty feet to the ground. Humphrey pulled away and fled, screaming. Figan chased him a short distance and then, still without pause for breath, climbed back into the tree and continued to leap about in the branches.

During the next fifteen minutes Figan displayed five more times. Twice he attacked a low-ranking male and the frantic screaming of his luckless victim added to the general confusion. Finally Figan became still (he must have been quite exhausted) and sat with heaving sides. Seeing this, Humphrey, who had unobtrusively climbed back into the tree, made himself another nest. Too soon! He had barely laid his head on a bunch of soft green leaves when Figan began yet another display and once again hurled himself down onto his rival. For a second time the two fell to the ground; for a second time Humphrey broke away and, screaming loudly, fled into the undergrowth.

By this time it was almost dark. Figan sat for a while on the ground and then climbed up the tree and made himself a nest. Only then did

Humphrey return and, very quietly, make his third bed. This time he was able to settle down for the night without further interruption.

Throughout that entire skirmish, big brother Faben had watched from his nest. I wonder if Figan would have dared attack his powerful adversary had Faben not been present? I suspect not. As it was, he surely knew that Faben would have helped him if he needed it. Perhaps more importantly, Humphrey knew it too.

After that decisive victory, a triumph watched by more than half the members of the Kasakela community, Figan's top rank seemed assured. But although he now accepted Humphrey's show of deference quite calmly, almost as his due, Evered, it seemed, was still perceived as a threat. After all, he had been dominant to Figan for years, and during his long quest for power had shown far greater persistence and vigour than had Humphrey. The grand finale came towards the end of May and, as before, Faben supported Figan throughout.

It took place on a hot, humid afternoon. The two brothers were feeding peacefully when Evered's distinctive pant-hoots sounded from the far side of the valley. They glanced at one another, their hair bristled, and they grinned widely in excitement. Then, leaping to the ground, they raced off in the direction from which the calls had come. They found Evered in a tree on a steep hillside. Terrified, he crouched there as the brothers charged back and forth below, dragging branches and hurling rocks. Then, as one, they leapt up into the tree and threw themselves on their victim. Locked together, grappling, the three males fell to the ground and Evered managed to break free. He fled some way up the hillside, then took refuge in another tree. The brothers followed and, for the next hour, displayed on and off below him. Poor Evered, there he stayed, occasionally whimpering and screaming in fear until, at last, Figan and Faben moved away. Not until they were some distance away and out of sight did Evered dare to climb silently from the tree and make his escape.

Figan had made it to the top.

A typical mixed group feeding on new shoots.

Prof resting.

Chimpanzees use more objects as tools for more purposes than any other creature except ourselves. The 'ant-dipping' technique (the ants have a vicious bite) as practised at Gombe.

Gremlin and Galahad termite-fishing. (Ken Regan/Camera 5)

Above: Flo looked older than any other chimpanzee I have known at Gombe. She was probably closer to fifty than forty. (B. Gray)

Below: Flo dangles Flint from her foot in her characteristic and unique way. (Hugo van Lawick)

Right: When Fifi is allowed to hold Flint, she imitates her mother. Subsequently Fifi was seen holding her own infants thus from time to time. (Hugo van Lawick)

Pom. The problems of adolescence. (C. Tutin)

Below: Passion is tense and nervous in her interactions with adult males. (P. McGinnis)

Opposite, top: Perhaps this is why Pom is nervous too. She sometimes dabs at a male's face in a peculiar way.

Old Flo peeks at her grandson as Flint grooms his sister. (Hugo van Lawick)

Figan with his powerful family. Left to right: Flint (in front of Fifi), Flo with infant Flame, Faben and Figan. The family utters pant-hoots in response to calls from other chimps. (P. McGinnis)

As Humphrey passes by, adolescent Figan shows respect, bending his arm and wrist and leaning away in a gesture of submission.

The charging display enables the male
to look larger and more dangerous
than he may actually be. He is often
able to intimidate his rivals without
recourse to aggression.
(Hugo van Lawick)

Faben, despite his paralysed arm, had
a magnificent display.
(Hugo van Lawick)

Evered displays.

Figan as alpha male.
(Hugo van Lawick)

6

Power

<<<<<<<<<<<<<<

I T IS ONE THING to rise to the top-ranking position of a community. It is another matter to remain on top day after day, month after month. Figan had attained his goal thanks to the support of his brother — and Faben would not always be around, for every hour of every day. How would Figan manage then if one of the other males should challenge the new order?

The test came all too soon when Faben, involved in romantic dalliance with a female, vanished for three whole weeks to the northern part of the community range. Figan was extremely worried — and rightly so, for Humphrey and Evered might well have challenged their new alpha had they realized that his ally was so far away. Every so often Figan would climb a tall tree and, from the higher branches, gaze out in all directions as through searching for signs of his missing brother. Occasionally he would give the long, loud screaming calls that serve to attract the attention of friends in times of need — SOS screams, we call them. But Faben was too far away to hear and Figan was forced to rely on his own resources.

It reminded me vividly of the time when, at the beginning of Mike's reign as alpha, we had removed his tin cans: for he had relied on them during his struggle for supremacy much as Figan had relied on Faben. In his effort to compensate for their loss Mike had expended huge efforts to make his displays impressive in other ways. He had hurled the very biggest rocks, dragged and flailed enormous branches — even two branches at a time. Once as he rushed towards a group of adult males with a palm frond in each hand, he had actually paused to gather up yet a third. Only very gradually had Mike relaxed, realizing that even without his precious cans he still held the respect of the other males.

And now, ten years later, Figan responded to a similar challenge in much the same way. The frequency and vigour of his charging displays increased dramatically, and he was a past master when it came to planning and executing these performances. Thus he would, if possible, move quietly upslope from some unsuspecting group, then charge down. Not only did this give him an element of surprise, but it enabled him to appear at his most impressive as he bore down upon the group, flat out, from above. And, of course, it is less tiring to run downhill; there will be more energy to spare if, in the face of any insubordination, it should be necessary to repeat the performance.

Most effective were his wild arboreal performances at the crack of dawn when it was still almost dark and the rest of the group was still abed. These caused pandemonium, with confused chimps screaming and hurling themselves from their nests. Back and forth, up and down — Figan leapt from branch to branch, shaking the vegetation, snapping great branches and, for good measure, pounding, from time to time, on some unfortunate subordinate. The confusion and the noise were unbelievable. And then, when it was all over, their new alpha, all bristling magnificence, would sit on the ground and, like some great tribal chief, receive the obeisance of his underlings.

And so, as a result of high motivation, determination and the expenditure of much physical effort, Figan stayed on top. And when Faben finally returned to the centre of the community range, Figan was able to relax and enjoy to the full the fruits of his labours — the respect of all the other members of his social group and the right of prior access to any feeding place or sexually attractive female that he fancied. Power.

One day, soon after Faben's return, I watched as the two brothers, who had been on their own for a while, approached three of the other males who were peacefully feeding on fallen fruits. As Figan, closely followed by Faben, charged towards them, all three screamed and rushed up trees. Their point made, the brothers sat with bristling hair and looked up into the branches above. Satan, a good deal larger than the new alpha, and in his prime, hastened down and, with loud pant-ing-grunts of submission, pressed his mouth to Figan's thigh. And Figan, utterly relaxed, utterly self-confident, laid a munificent hand on the bowed head before him. Then, as Satan began to groom Figan, Jomeo and Humphrey also approached to pay their respects and, for a while, Figan was groomed by all three.

Faben, probably because of his paralysed arm, had never become a high-ranking male. But as brother of the alpha he was treated with a new respect by the other males — at least when Figan was around. Faben probably realized this quite quickly, for, after that initial three-week period in the north, he rarely spent more than a few days away from Figan.

Some adult males spend a good deal of time on their own — Mike, even when alpha, had sought occasional spells of solitude. But Figan, from earliest childhood, had wanted to be in the thick of things, been happiest when part of a noisy, excitable group — males, females, the more the better. And Faben, now that he was spending so much time with Figan, became more social too. The two brothers formed, in a way, the hub around which the wheel of society revolved. The other chimps, particularly the males, were fascinated as well as intimidated when Faben, charging along with his splendid upright gait, limp arm swinging, hair bristling, joined in the already impressive displays of their alpha.

For the first two years of his reign Figan held a position of almost absolute power in the community. This meant that he could, if he so wished, maintain all but exclusive mating rights over any female who caught his fancy. Once he had proclaimed his interest by threatening any would-be suitors who approached too closely, his mere presence, close to the lady friend of the moment, was usually sufficient to inhibit the sexual advances of the other males. He established a pattern, taking over the community females, one after the other, when they were at their most alluring — during the last four or five days of their swell-ings.

Faben's privileged position was very apparent at such times for Fi-gan usually shared his sexual possessions with his brother much as he shared precious food items, such as meat. And Figan received a payoff for his generosity: Faben helped to keep an eye on the current lady friend when Figan was momentarily busy elsewhere. However, even Figan and Faben between them could not prevent their female from enjoying occasional clandestine intercourse with one or other of the frustrated lower-ranking males. Such opportunities arose when the attention of the alpha male and his brother was temporarily diverted. Once, for example, when Figan and Faben were intently watching a troop of colobus monkeys with an eye to acquiring monkey meat, three other males copulated with their female in quick succession: neither of the brothers even noticed!

It always surprised us that the females themselves were prepared to cooperate in these illicit affairs. Because when Figan did notice he would race towards the pair and, very often, bash the female for her faithlessness. This made more sense than attacking the rival male — for a skirmish of that sort would have left the female unguarded and available for yet another quick clandestine mating!

The male who sneaked the most copulations with Figan's females was adolescent Goblin. He was utterly fascinated by sex and, incidentally, utterly fascinated by Figan, too. Because he was not perceived as a rival (he was only nine years old when Figan came to power) Goblin was able to maintain surprisingly close proximity to the succession of females with whom the alpha male satisfied his sexual needs. Thus, even if Figan's attention was diverted but momentarily, Goblin was on hand to take advantage. And since the sexual act comprises no more than ten to twelve rapid pelvic thrusts, the briefest of opportunities sufficed — so long as the females cooperated, and for some reason they usually did. So closely did Goblin follow those tempting pink bottoms that he was occasionally able to snatch a few seconds of sexual gratification as Figan led the way through dense undergrowth.

Sometimes an adolescent male selects one of the senior males as his 'hero'. He is attentive to all of them, but it is his hero whom he watches most closely, and with whom he is most likely to travel when he leaves his family. Figan, without a shadow of doubt, was Goblin's hero. Often, after watching Figan closely, Goblin imitated his behaviour. One day I watched as Figan did a magnificent display, dragging a large branch, slapping and stamping on the ground, and drumming on the buttress of a large tree. Goblin, from a discreet distance, watched intently and then displayed in his turn, following the exact route that Figan had taken, dragging the self-same branch and drumming on the same tree. I was reminded of those times when Figan had practised with Mike's empty cans.

Figan, for his part, was remarkably tolerant of his small and persistent shadow, but very occasionally, when Goblin got too close — when he was feeding, for example — Figan threatened him mildly. This would throw Goblin, temporarily, into a frenzy of apology. Sometimes Figan supported his young friend if he got into trouble with other individuals. Little did any of us realize then the far-reaching consequences, both for Figan and for Goblin, of this special relationship between them.

Under the rule of a powerful male the conflicts between the other members of the community are kept to a minimum, for he uses his position to prevent too much fighting among his subordinates. What motivates him is not always clear. Sometimes there may be a genuine desire to help the underdog. At other times it may be that the alpha feels his position is challenged if another male initiates a fight. I remember once when Figan and Faben jointly attacked a female during the excitement of a reunion. But when, a few moments later, young Sherry attacked the same female, Figan, a picture of chivalry, raced over, bashed the aggressor and so 'rescued' the female. But whatever the driving force behind Figan's interventions in the affairs of his underlings, his behaviour served to terminate countless squabbles. Moreover, I suspect that many would-be aggressors, anticipating the displeasure of their boss, exercised more self-restraint when he was around. Thus Figan, during the years of his power, helped to promote and maintain an atmosphere of social harmony among the members of his group.

During the second year of Figan's reign two of the students — David Riss and Curt Busse — asked me if they could follow Figan, monitor his movements, behaviour, and relationships with other chimpanzees, for fifty consecutive days. I was not sure. Perhaps this would be too much of an intrusion into his life, make him uneasy or irritable. But there was a precedent — six years earlier Flo had been followed for sixteen days in an attempt to witness the birth of her last infant (the attempt failed as the baby was born at night). Flo had not appeared to mind at all, and Figan was as tolerant of humans as she had been. And so I agreed — on condition that the follow be called off if Figan became upset.

The marathon began on 30 June 1974 and continued until 18 August. David and Curt, each accompanied by one of the field staff, spelled one another every four days, so that while one of them was clambering around the mountains after Figan the other was writing up the information he had collected — and resting after the arduous four days of following. The fifty days with Figan gave us invaluable data about the behaviour and social life of one of the most powerful top-ranking males Gombe has known, at a time when he was at the zenith of his career.

In those days, when all the students gathered together for dinner, there was a great exchange of information every evening. Many were the tales told around the tables in the mess. There were Caroline Tutin's accounts of the sex life of the various females, Anne Pusey's descriptions of adolescence, Richard Wrangham's stories of feeding and ranging behaviour, and countless anecdotes concerning the development of infants recounted by the various young people involved in the long-term mother–infant study. And now we had, in addition, daily reports on Figan.

During the fifty days there were two sexually popular pink females, and Figan monopolized them one after the other. The first of these was Gigi. Large and sterile, Gigi, who has shown one sexual cycle after another since 1965, uninterrupted by pregnancy and childbirth, is, in many ways, rather masculine. She has a mind of her own and does not submit readily to male bullying. There was no doubt but that, during the days when she was fully pink, she controlled Figan's movements and thus those of his entire group. One day, for example, when the chimps were headed towards a stand of *kifumbe* fruits, Gigi suddenly left the trail and plunged into the undergrowth. Figan and Faben followed at once, while the others hung about, waiting. Some climbed to feed on other fruits nearby, the rest sat or lay on the ground.

Gigi made for a nest of *siafu* — those vicious, biting driver or army ants that are such a delicacy for the chimpanzee. Upon arrival at the site she broke a long straight branch from a nearby bush, removed the side branches, then carefully stripped the bark until she had made a smooth tool, about three feet long. She reached her hand a short way into the opening of the nest and, for a few seconds, dug frantically until the ants began to swarm out. Quickly she plunged her tool into the nest, waited for a moment, then withdrew it covered by a seething mass of ants. With rapid movements she swept the stick through her free hand, pushed the ant-mass into her mouth, and crunched vigorously. As the ants poured out of the nest in ever greater numbers, agitated by the intrusion, Gigi climbed a sapling nearby and, reaching down with her stick, continued her meal. Every so often she had to slap frantically at her feet and kick at the trunk to repel those ants that were finding their way to the source of the raid. Now that she was using one hand to hold onto the sapling while she fished with the other, she had to transfer the tool to one foot between each dip, thus freeing a hand for sweeping the ants into her mouth. Nevertheless, despite these difficulties, she persisted.

Figan, meanwhile, had begun to fish for *siafu* as well. But after only ten minutes he left his tool and rushed away to pick off the ants that had crawled up his arms and legs. Faben then picked up the abandoned tool but after fishing for only a couple of minutes he too gave up. The two brothers then started off in the direction of those delicious *kifumbe*.

Gigi, however, did not follow. She had positioned herself, by this time, on a low branch directly above the nest and, from this place of comparative immunity, continued to feed on ants. So Figan and Faben sat and waited. After a while Faben lay and closed his eyes. But Figan gradually became more and more impatient. Seven times he uttered his characteristic 'Let's go!' grunt, but Gigi completely ignored these pleas. From time to time he shook little branches at her, requesting that she should follow him. But he did not do this very vigorously, and she paid absolutely no attention. Only when she had been fishing for forty-five minutes (with an average of about two stickfuls of ants per minute) did she finally give up and join Figan. Then the three of them moved after the rest of the group.

The following day, when Gigi's feeding preferences conflicted with his, Faben left her and went off with the group. But Figan remained faithful. For a total of one hour and twenty-minutes, spread over five different episodes during the day, he waited patiently while she fed, grumbling his soft 'Let's go!' grunts from time to time. But only when she had quite finished feeding did she climb down and calmly follow where he led. By the following morning Gigi's swelling had waned and Figan's proprietary interest in her ended.

During those few days when Figan and Faben were both dancing attendance on Gigi, one most unusual event took place, when Curt was following them:

'Just after they'd left their nests I saw Faben mating Gigi,' he told us that evening. 'Suddenly Figan noticed and charged at them with his hair out. He actually stamped on Faben's back. He stamped three times, quite hard, and Faben screamed like anything, and then waa-barked as Figan charged off. Just a bit after that, Figan mated Gigi himself.'

'That's about the only time Figan has minded sharing his female with Faben, isn't it?' I asked.

'I saw it happen one other time,' said Caroline. 'That was when Faben was mating in thick bush — I don't think Figan realized who it was for a few moments. They both looked surprised afterwards!'

When Patti went pink in her turn, Figan did not make any obvious attempts to prevent Faben from mating with her. And after she had subsided there were no more pink females for the rest of the fifty-day follow. It would be crude and altogether disrespectful to an alpha male to describe here David's observation, made six days after Patti's de-tumescence, which led him to suspect that Figan, sound asleep in his nest, was dreaming about the sexual delights of the previous weeks!

One evening Curt had an exciting story to relate. Figan, travelling with Faben, Satan, Goblin and four females, had begun hunting ba-boons. While Faben and Goblin sat below and watched, Figan had climbed slowly towards a baboon mother and her small, black infant. But she was alert and, although he chased her a short way, she easily escaped.

'Do you know who it was?' asked Tony Collins, one of the students studying baboons.

'Yes. It was that A troop mother with the blind infant — what's her name — Hokitika isn't it?'

'Well, I'm glad she escaped,' said Craig Packer, another member of the baboon team. We were all glad, although the future for a blind infant baboon was hardly rosy and, in fact, she died just one week later.

After that, Figan had remained up the tree a while, looking in all directions. Suddenly he had climbed to the ground and hurried lower down the slope. As he approached a tall, dead tree — little more than a post, with stumpy, broken-off branches — he had begun to move cautiously and silently. Peering through the foliage, Curt had seen, up near the top where the dead tree was thickly draped in vines, a very small baboon — little more than an infant. There was an adult male baboon feeding some thirty yards away, but he had taken no notice as Figan slowly climbed towards his intended victim.

'Fig suddenly made a rush towards the infant. He nearly caught it, too. But somehow or other it escaped and leapt to the ground. It was amazing — it must have been at least forty feet, that leap. And then the little thing landed right between Faben and Goblin!'

'Now I suppose you're going to describe a horrible, gory kill,' said Julie Johnson, another of the baboon team. 'I don't think I want to stay and listen.'

'No, it was okay,' Curt reassured her. 'Just at that moment the male baboon finally arrived and there was a great commotion. The little bab got away. The male pitched into Goblin and there was a truly spectacular fight. I don't know how Gob did it, but somehow he won and chased off after the bab. And just at that moment another big male arrived. We knew him — it was Bramble. He began to threaten Faben, and two female baboons joined in. Faben was quite scared and rushed up a tree.'

'Didn't Figan help him?' I asked.

'No — he just sat and watched. In that same place where he'd almost caught the baby. Then, after a bit, he climbed down and the chimps all wandered away.'

In fact, Figan and his group hunted relatively seldom during those fifty days. There were eight colobus monkey hunts and seven kills —

Figan, who has always been a successful hunter, made three of the kills himself.

Nor did they make many journeys to the peripheral part of their home range. Once they travelled far to the south, penetrating the overlap zone between their community and the neighbouring Kahama community. They heard calls that were presumably made by Kahama chimpanzees and became very excited, embracing one another, grinning, travelling silently, and spending some time gazing southward from a high ridge. But nothing further happened, and presently they all returned to the north, displaying frequently and calling loudly, as though to release the tensions that had built up while they were close to strangers.

Figan, as might be expected, spent more time with Faben than any other adult male, and young Goblin was often tagging along with them. Figan also spent many days with Gigi, not only when she was pink, but also when she was flat and sexually uninteresting. And quite often he was with his sister Fifi and her infant son Freud. Most of his interactions with the individuals of this community were, at that time, relaxed and friendly. He was so clearly dominant over them that, except when there were moments of tension such as during a reunion, he had no need for violent demonstrations of strength and mastery.

Unless Evered was around. And then Figan, joined almost always by Faben, displayed with unusual frequency and vigour. It was as though despite his position of great power, despite the support of his brother, and despite the memory of those clear-cut victories over Evered the year before, Figan still felt threatened by the rival of his adolescent days.

David was bursting with excitement one evening when, as usual, we had all gathered in the mess.

'I saw the most unbelievable attack on Evered today,' he said. 'The whole thing lasted for almost two hours.'

It happened when Evered, by himself, joined the group. He didn't see Figan and Faben at once, for they were feeding in thick undergrowth. But suddenly they charged towards him and he rushed, screaming, up a tree. Figan and Faben displayed below him a few times, then they settled down on one of the lower branches of his tree and began, very calmly, to groom one another.

'It was pathetic,' said David. 'Evered was about twenty feet above them and he whimpered and gave little screams almost non-stop. He was watching them all the time, but they just ignored him and went on grooming.

'After that,' David went on, 'Figan and Faben left the tree and did some more fabulous displays. They charged about together — four times in the next half hour.

'Then came the actual violence. Figan started it — he went leaping up into Evered's tree and kept chasing him from branch to branch. After a bit Evered managed to leap to another tree, but Figan followed.

'And all the time Faben was charging about on the ground below and Evered was screaming, terrified out of his wits, and keeping as far as he could from Figan.'

David paused. 'It was awful really, watching it all,' he said. 'It was almost like seeing a cat playing with a mouse, because I knew that there wasn't any way that Evered could escape — unless they actually allowed him to.'

By this time we were all caught up in the drama, tense and expectant.

'Suddenly Evered made a huge jump into a third tree,' David continued. 'Figan leapt after him and Faben suddenly rushed up as well and they had Evered sort of stuck between them. And then they both jumped him together, and they all fell and just went on fighting on the ground till poor old Ev got away.'

'Poor old Ev' it was, for the brothers followed, and again cornered and attacked him. He managed to get into a tree and his persecutors continued to charge about in great excitement for a further ten minutes until, perhaps because another adult male arrived on the scene, Figan and Faben left and Evered, still screaming, was finally able to escape.

A month later Figan and Faben encountered Evered after a two-week separation. Curt observed the reunion which took place in a tall tree. It was tense and dramatic. Figan and Evered embraced, both screaming. The other chimps present were watching intently. They too were highly excited and screaming loudly.

'I was looking up, doing my best to see exactly what was happening,' said Curt, 'when the unimaginable happened.' He paused dramatically and we all wondered what was coming next. 'Well, you know what fear and excitement can do to your guts,' Curt went on. 'One of those wretched creatures — I'm pretty sure it was Gigi — suddenly let go. I was absolutely showered with warm shit!'

Of course we were sorry for him, but nevertheless the whole mess collapsed laughing while Curt tried to look pained and aloof. Poor Curt — he had had to leave all the excitement and go and wash off in the stream. He was lucky that there *was* a stream close by! Fortunately he was with Eslom, who had recorded the details of the fight that took place.

On that occasion Evered was set upon by five aggressors, for Humphrey, Gigi and an adolescent male had joined forces with Figan and Faben. The attack looked — and sounded — incredibly violent and it was amazing that Evered sustained only a few small wounds. He stayed with the group for the rest of the day, but left before the others settled down for the night and was not seen again for another two weeks.

It was hardly surprising that, in the face of this bitter persecution, Evered spent less and less time in the central part of the community range. It really seemed as though Figan, with Faben's help, was actually trying to drive Evered right out of the Kasakela community.

And then, quite suddenly, things changed. Almost exactly two years after he had taken over as alpha male, Figan's days of absolute power came to an end. Faben disappeared — this time for good. Gradually the other males must have realized what had happened for they began to capitalize on Figan's vulnerable position. In groups of two, three or more they ganged up on their alpha in dramatic confrontations. It seemed that he could never hold his own against them.

But by that time, in June 1975, there were no longer any American or European students at Gombe to record the events.

7

Change

I N M A Y 1975 came a sudden night of terror: forty armed men came across the lake from Zaire and kidnapped four of the Gombe students. Afterwards there were many confused tales of what had happened, tales of courage as well as tales of horror. My old friend Rashidi was beaten over the head in a vain attempt to make him reveal the whereabouts of the key to the petrol store. He was deaf in one ear for months afterwards. The two young Tanzanian women working at Gombe then, Park Warden Etha Lohay and student Addie Lyaruu, flitted from one student's house to the next, moving quickly through the dark forest, to warn everyone of the attack.

Where had the victims gone? Were they even alive? There were reports of gunshots heard out on the lake, and for days we thought that the hostages might have been killed. It was a time of anguish. Of course we all had to leave Gombe. For a while we stayed in Kigoma, hoping against hope for news of our friends. But none came. A few months before the kidnap I had remarried, and my second husband, Derek Bryceson, had a house in Dar es Salaam. There we all went, the students crammed into the little guest house, and there we waited. Waited and waited and waited, for what seemed eternity, for news. If it was pure hell for us, those who had not been taken, what of the mental suffering of the victims themselves, and of their parents and other close family?

After about a week, which seemed like a month, one of the kidnapped students was sent back to Tanzania with a ransom demand. I shall never forget the relief, the delirious joy, that I experienced on learning that the four were alive and at least physically unharmed. But the negotiations seemed to go on for ever. The issue was highly sensitive politically, involving as it did relations between Tanzania, Zaire and the United States.

It was fortunate that all four of those young people were mentally as well as physically strong, and fortunate too that they had each other for moral support. Perhaps the worst anguish was during the final days, when one student was kept behind, a lonely hostage, after the ransom had been paid and the others released. But after another two weeks he too was released. It was as though a black cloud had finally moved aside and allowed the sunlight to come flooding back.

All four eventually recovered from the terrifying ordeal — at least they seemed to have, judging by outward appearances. But I wonder if they will ever entirely free their minds from the psychological torment of those days. The memory, surely, will always be lurking there, ready to erupt in nightmares in times of sickness, loneliness or depression.

During the period between the night of the kidnapping and the final release of the last hostage, my thoughts of the research at Gombe had been stifled, crushed beneath the load of worry and despair. For a while I had organized some analysis of the data, something to try to keep up the morale of our little group in Dar, but our hearts were not in it. Most of the time I just read novels — I hadn't read so many novels since my school days. But once the hostages had been released I could think again about the future of the research. Derek, Grub and I had made several brief visits to the park, even during the nightmare weeks. We had to encourage and show support for the field staff who, to their great credit, had continued to record basic data entirely on their own initiative.

Immediately after the raid, a detachment of the Field Force, a special branch of the police, had been sent to Gombe. This highly efficient force, trained to handle all emergencies, was a great comfort to us during our early visits. After a few months it was replaced by a small group of ordinary policemen. Very gradually a feeling of security returned. When we visited, we no longer wondered whether we should take to the forest every time we saw an odd-looking boat. But it was more than a year before I could hear a motor boat stop in the night without leaping up, heart pounding, to gaze toward the lake, wondering whether we should flee up the mountain side.

Without Derek's help and support I doubt that I could have maintained Gombe after the kidnapping. I had met him in 1973 during a visit to Dar es Salaam and we had, immediately, felt a strong attraction. He had first arrived in Tanzania in 1951. During World War II he had been a fighter pilot in the RAF, but after only a few months of active service, had been shot down in the Middle East. He survived the crash but suffered a spinal injury, and was told he would never walk again. He was nineteen years old at that time. Determined to prove the doctors wrong, he had, through sheer determination, taught himself to move about with the aid of a stick. He had just enough muscles in one leg to move it forward as he walked: the other had to be swung forward from the hip. He learned to drive too, fast and well, even though he had to lift his left leg with one hand in order to transfer his foot from clutch to brake!

Once mobile, Derek had gone up to Cambridge, where he acquired a bachelor's degree in agriculture. He was then offered a job in England which he instantly rejected. 'It was cushy armchair farming,' he told me, 'suitable for an invalid.' Instead he raised the funds to get himself to Kenya where he farmed for two years, then applied to the British government for one of the beautiful farms on the foothills of Mount Kilimanjaro in what was then the British protectorate of Tanganyika. There he became a successful wheat farmer — until he met Julius Nyerere, who was then organizing the movement that would eventually lead to Tanganyika's independence. Derek was deeply impressed by Nyerere and became sympathetic to his cause. This changed the course of his life. He joined the Tanganyika African nationalist movement and became so involved in politics that he gave up his beloved farm and moved to the capital, Dar es Salaam. Thus he was firmly entrenched in his adoptive country's politics when independence was finally won in 1961 — just after I arrived at Gombe.

Derek did much for Tanzania — as Tanganyika became after its union with the island of Zanzibar. He was elected member of parliament for the huge Dar es Salaam constituency of Kinondoni, and was returned, with landslide majorities, every fifth year. He held many cabinet posts, but was best known for his contributions to Tanzanian agricultural policies during two five-year terms as Minister for Agriculture, and for his development of preventive medicine programmes and improved standards of nutrition during the years when he was Minister for Health. When I met him he had resigned from government, but still represented Kinondoni as a member of parliament, and had recently been appointed director of Tanzania's spectacular wildlife parks by President Julius Nyerere.

After Derek and I were married, I had continued to live at Gombe and he had made periodic visits, flying in for a couple of days at a time in a four-seater single-engine Cessna. Derek loved to watch the chimps, but it was not easy for him to climb the steep slope to camp. We cut steps into the steepest, most treacherous parts of the trail, and rigged up a rope in the very worst stretch so that he could support himself with this on one side while using his stick on the other. This allowed him to go up and down by himself, without leaning on a friendly arm as he had had to do before. But even so, the journey that took the rest of us about ten minutes was a forty-five minute endurance test for him. Once he slipped and landed heavily on the tip of his spine and was in great pain — though he would never admit it — for several days. Another time he fell and wrenched his knee which swelled to a tremendous size. But despite the hazard he always insisted that it was worth it.

During those visits Derek, as director of national parks, had made it his business to become conversant with all that was going on at Gombe. Thus, after the kidnapping, he was able to be really helpful. With his fluent Swahili and his understanding of the Tanzanian character, he helped me to convince the field staff that they could do good work on their own. Although they had acquired so much knowledge and experience during the preceding few years and could follow the chimpanzees skilfully through the forested mountainous terrain, chart their daily movements and association patterns and identify their food plants, they had come to rely on the guidance of the students and the constant presence of 'Dr Jane'. Now it was necessary to convince them that they could carry on without us.

I worked with the men closely during my all too brief visits, checking on their accuracy and reliability. We gathered together for talks and seminars and I told them about the analysis I was doing in Dar es Salaam — for I had begun to pull together the results of the study for eventual publication in a scientific book. When they understood how I would use the information they were gathering they took more care in writing their reports, making out their charts and maps. Gradually their confidence grew. They elected, from among their number, two *viongozi* or leaders — Hilali Matama, who had begun work with the chimps in 1968, and Eslom Mpongo, who had joined our team soon afterwards. By 1975 the two of them knew as much about chimpanzees and their behaviour as any so-called 'expert' — and more than most. Their work had become a way of life, and they, and the other members of the team at Gombe, were utterly absorbed and fascinated by the lives of the chimpanzees they were observing. Each time I returned to Gombe I taught them to collect ever more sophisticated data and their reports became increasingly rich. We provided them with a tape recorder so that, if they chanced to witness some exciting or unusual event, they could dictate a more detailed report than they could have put down on paper. Most of them wrote rather slowly and laboriously — one or two, in fact, had only recently learned to write in order to join our staff.

The Tanzanians worked in teams of two, following a selected 'target' chimpanzee for as long as possible during the day — ideally from the time when he or she left the night nest until nightfall. One of the men recorded, in detail, the behaviour of the target. The other plotted the travel route, listed the foods eaten, and kept note of the other chimpanzees who were encountered and how long they remained with the target. Between them the men also described any interesting events concerning individuals other than the target. Often, after supper, the two men who had been out following would come to tell us what they had seen during the day. We would sit companionably on the soft sand outside the house, with the waves lapping or slamming into the shingle, and listen to the musical Swahili voices describing a hunt, a boundary patrol, or some amusing incident they had observed.

Each of the men had his own individual interest. For Hilali it was the male dominance scene. What a lot he — and the other men — had to tell us during the troubled months after Faben's death when, with increasing frequency and enthusiasm, the other males ganged up against Figan. It quickly became apparent that Figan, having relied, throughout his life, on the support of a close ally (first his mother, then his brother) found it necessary to cultivate a substitute for Faben. He chose Humphrey, his erstwhile bitter rival. It made good sense for, of all the males, Humphrey had been most terrorized by Figan and had suffered the greatest defeat. Thus he posed the least threat now. And, while he could not take Faben's place — for he never actively supported Figan when the other males challenged him — he provided a measure of comfort as he almost never ganged up with the others *against* Figan.

One evening in March, some eight months after Faben had vanished, Hilali arrived at the house eager to tell us about his day. He had been following Figan who, as usual, had been part of a large group. During a sudden outbreak of excitement as Satan joined them, four of the adult males — Satan himself, along with Evered, Jomeo and Sherry — had ganged up on their alpha in a series of dramatic joint displays. Three times, in the space of forty minutes, the four charged at and around Figan, causing him to flee, screaming. Eventually he took refuge in a tall tree, but the four followed him to the topmost branches. Terrified, Figan leapt wildly to a neighbouring tree, hurtled to the ground, and ran as though pursued by the bats of hell, for at least five hundred yards. Hilali, exhausted and with the sweat pouring from his brow, somehow managed to keep up, and thus saw Figan, still screaming loudly, leap into a tree and fling his arms around Humphrey. Hilali thought that Figan had probably seen his one ally from the tall tree — although it might have been a fortuitous meeting. The other four males continued to display at both Figan and Humphrey, who stayed very close together, each seeking reassurance from the other.

Many similar incidents were reported during those tumultuous months when the relationship between the adult males was so strained and tense. And always Humphrey, when present, provided moral support for Figan. The extent to which Figan came to rely on Humphrey was well illustrated during one of Hamisi Mkono's follows. During a feeding session in thick undergrowth, the two friends became temporarily separated. When Figan suddenly realized that Humphrey was no longer with him — '*alianza kulia kama mtoto*,' Hamisi said, laughing — he started to whimper like a lost child. He climbed a tree, staring out in all directions, and then hurried off to search for his friend, every so often screaming — those SOS screams — at the top of his voice. After about twenty minutes he found Humphrey, rushed up and begun to groom the older male. Gradually he calmed down.

I think we all expected that Figan would lose his alpha position for good. Indeed, for about nine months there was no clear-cut alpha male at Gombe. Figan could — and did — hold his own when he encountered the other males on their own, or in pairs. But he ran from them, screaming, when they ganged up in groups of three or four. What was it about him, I still wonder, that prevented the other males from following up their advantage, on such occasions, and joining forces to actually attack Figan? They never did. And most of the dramatic confrontations, the bristling charges and wild swaying of vegetation and hurling of rocks, ended with all participants suddenly rushing together, screaming, and starting somewhat frenzied sessions of social grooming — during which all concerned gradually calmed down and, after a while, moved off together.

It was during this troubled period that the sexually popular female, Pallas, came into oestrus again after losing an infant. And, with no clear-cut alpha, this caused almost total chaos among the males. Figan no longer had the power to take sole possession of a hot number like Pallas — nor had any of his rivals. And so, almost every time one of the big males climbed her tree (for, probably in sheer self-defence, she spent most of her time above the ground), pandemonium broke out among the others. Either the daring suitor himself was chased up the tree and attacked by one or more of the other males, or, if he made it to his goal, the sight of the sexual act triggered aggressive outbursts among the spectators. And then there would be a brief spell of bedlam as males displayed with bristling hair and furious scowls, hurling rocks and occasionally seizing and pounding some luckless female or adolescent who got in their way. Sometimes they engaged in brief but furious battles between themselves. Pallas herself was rarely a victim, but she must, nevertheless, have suffered through any number of almost unbearably tense moments.

Throughout this incredible ten-day period, Goblin — who, incidentally, continued to follow Figan faithfully, despite the temporary dethronement of his hero — stuck close to Pallas through thick and thin. Sometimes he was attacked for his audacity, but he got in many quick copulations while his elders fought each other for the privilege of access.

After nine months of tension and anxiety, Figan once again established himself as alpha — though his days of absolute social power were over for good. And just as Faben had benefited from his status as brother to the alpha male, so now did Humphrey from his position as 'best friend'. Hilali recorded one delightful example when Figan — who was Hilali's favourite chimp — caught two infant red colobus monkeys during the same hunt. He found the first one almost immediately, grabbing its mother, pulling the baby from her arms, and killing it with a quick bite into the skull. And then, instead of starting to feed, he just sat, holding the limp body of his prey in one hand, intently watching two of the other males who were still hunting. After a few moments Humphrey climbed rapidly towards Figan and sat close to him. Humphrey was not interested in the ongoing hunt — only in begging from the share of Figan's prey. All at once, to Hilali's amazement, Figan thrust the entire carcass into Humphrey's hands. Then, leaping from the tree, he raced to rejoin the hunt and, within a few minutes, had got hold of a second mother, seizing and killing her infant. This time he consumed his prey himself!

'*Ni fundi, kweli!*' — he's truly an expert, said Hilali, chuckling. He stared into the fire for a moment and then, as though feeling the need to be absolutely fair, to give credit where credit was due, he added: '*Na kumbuka Sherry, anapofanya hivyo*' — I remember Sherry doing the same. Indeed, Sherry had, in a way, gone one better: he had caught a second prey while still clutching the better part of his first kill. And he kept and consumed them both!

Throughout the early post-kidnap years Derek continued to help with the administration and organization of the research at Gombe, and, as the months went by he seemed to get busier and busier. To all intents and purposes he had two constituencies, each with its own pressing needs and problems: the Kinondoni district of Dar es Salaam, whose inhabitants he had represented in government for nineteen years, and Tanzania's national parks, whose furred and feathered inmates were equally in need of his political skills and wisdom. The non-human occupants of the Gombe national park, safe in a highly protected environment, needed his help less then most others, so that it became increasingly difficult for him to justify more than an occasional and very brief visit to see the chimpanzees that he loved.

By that time, however, it was considered safe for me to go to Gombe on my own. When Grub (who had, for a while, continued 'school' in Dar es Salaam, doing his lessons in a little room next to my office) went off to a prep school in England, I was able to spend more and more time there. It seemed strange at first, with just the Tanzanians and myself — more like the early days when I had spent months at a time with Hassan, Dominic and Rashidi for company. I missed the students — for a while, indeed, I felt that it would be impossible to keep Gombe going without them. But as the months went by I gradually adjusted to the new state of affairs and I found that the pattern of my life — living in Dar es Salaam with visits to Gombe as frequent as I could make them — had some decided benefits. When I was at Dar es Salaam I could concentrate on analysing and writing. I set up a breezy office where the data could be stored, and where I could work at my desk and gaze out over the bougainvillea — an exotic riot of colour, purple and pink, crimson and orange-yellow, white and green — to the deep blue of the Indian Ocean. And when I was at Gombe I could throw myself into working with the chimpanzees, following them through the forests, immersing myself in their lives.

Even during the days when I was away from Gombe, Derek and I maintained close contact with all that was going on there, speaking to the men daily by two-way radio. It was over the radio that we heard, one morning, that Gilka had given birth. I was delighted, for her first baby had mysteriously disappeared when he was just under a month old. But my joy was short-lived: three weeks later another radio message about Gilka, distorted and indistinct, brought horrifying news from seven hundred miles away. Indeed, Derek and I found it hard to believe: '*Passion amemwua na amemla mtoto wa Gilka*' — Passion has killed and eaten Gilka's infant. Derek turned off the radio and looked at me.

'It can't be true. It can't,' I said. And yet I knew it must be. No one could invent such a horrifying incident. 'Oh!' I burst out, 'why, why, why did it have to happen to *Gilka*?'

8

Gilka

◀◀◀◀◀◀◀◀◀◀◀

THE CAREFREE DAYS of Gilka's life ended when she was about four years old. As a small infant Gilka did not lack for companionship: her elder brother Evered was usually around, and her mother, Olly, spent much time with Flo and her family. But Evered was eight years older than Gilka — presumably Olly had lost at least one infant between the two of them — and he began to leave his family for extended periods when his sister was only five years old. At about the same time Olly began to avoid Flo because Figan, entering adolescence, not infreqently challenged his mother's friend with blustering displays. And so Gilka spent hours, sometimes days at a time with only her timid mother for company. How happy for her we were when her infant brother was born. Soon he would be old enough to play with her and her days of loneliness would be over. But then came the grim days of the 1966 polio epidemic when Olly's month-old infant became sick and died, and Gilka herself was particularly paralysed in one wrist and hand. Then, as though all this was not enough, two years later Gilka developed a bizarre fungus infection which, by the time she was eleven years old, had hideously disfigured her once elf-like, heart-shaped face. The grotesque swellings on her nose and brow ridge spread to her eyelids so that she could barely open her eyes.

Once we had diagnosed the disease we were able to control the symptoms with medication. But when Gilka transferred, temporarily, to the community in the south we were not on hand to lace her bananas with medicine and she returned after six months all but blind. (She may have been pregnant, too, but if so she lost the baby.) Once again we were able to bring the swelling under control and soon, to the obvious satisfaction of the adult males, she resumed her interrupted periods of sexual swelling. Gilka, like the majority of adolescent females, enjoyed sexual interactions but she often had difficulty in keeping up with groups of fast-moving males because her bout with polio had wasted the muscles of her left arm. Although I suspected that she was somewhat relieved whenever her exhausting pink days were over for a while, she was, nevertheless, a lonely chimp in between periods of sexual activity — her mother, old Olly, had died by this time and although her relationship with Evered was still excellent, he was not often around to keep her company.

Then, in 1974, things seemed to change for the better. Gilka appeared one day with a tiny infant. We named him Gandalf, and hoped that his mother's days of loneliness would now be over — for once a female chimpanzee starts a family she seldom spends any time alone for the rest of her life. Moreover, the birth of a female's first baby often seems to induce an added respect for its mother among the other members of a community, male and female alike. It was wonderful to see Gilka, who had so often sat on the outskirts of any grooming or resting group, at last taking a more active part in community life. The arrival of this baby did one more thing for Gilka: after his birth we had decided to discontinue the medication for his mother's fungus infection, fearful that it might harm her baby. The swelling, rather than worsening, as we had feared, was instead noticeably reduced. After a while Gilka was left simply with an enlarged nose that was almost comic.

Gilka was an attentive and careful mother, just as Olly had been, and Gandalf, by the time he was a month old, seemed a healthy and well-developed infant. And then he vanished. We had no idea what might have happened — Gilka simply appeared one day without him. Once more, except during the days when she was pink, she began to wander about alone. And her fungus condition worsened.

It was almost exactly a year after Gandalf's disappearance that we received the radio message that Gilka had given birth again. The baby was a female, and we decided to call her Otta — planning to put an O back into the family names to keep alive Olly's memory. This was the infant who was killed by Passion.

When Derek and I got to Gombe we heard the horrific story in gruesome detail. Gilka, we were told, was sitting peacefully in the afternoon sun, cradling her tiny infant, when Passion suddenly appeared. She stood for a moment, looking at mother and child — then charged towards them, hair bristling. Gilka fled, screaming, but she was doubly handicapped — with an infant to support and a crippled wrist. In a flash she was overtaken. Passion leapt upon her and seized hold of little Otta. Gilka tried desperately to save her baby, but she had no chance and after the briefest of struggles Passion succeeded in snatching Otta away. Then, most macabre of all, she pressed the stolen baby to her breast, and Otta clung there desperately while Passion again leapt on Gilka. At this moment Pom, an adolescent at the time, rushed to join her mother, and Gilka, outnumbered, turned and fled with Passion in hot pursuit, Otta still clinging tightly to her belly. Confident in her victory, Passion sat on the ground, pulled the terrified infant from her breast, and bit deeply into the front of the little head: death was instantaneous. Slowly, with utmost caution, Gilka returned. When she was close enough to see the limp and bleeding corpse she gave a single loud, bark-like sound — of horror? despair? — then turned and left.

For the next five hours Passion fed on Gilka's baby, sharing the flesh with her family, Pom and juvenile Prof. Between them they consumed it all, every last scrap.

We were all dumbfounded. It was not the first example of cannibalism at Gombe — five years earlier a group of adult males had come upon a female from a neighbouring community, attacked her savagely, and during the fight had seized her baby, killed it, and eaten part of the little body. But that was different for the female was a stranger, an alien, who had aroused the hostility of the males. They had attacked her as part of their constant effort to protect their territory from intrusion by outsiders and her infant, it seemed, had been killed almost accidentally. Only a very small portion of the body had been eaten, and by only a couple of the males present. For the most part the aggressors had displayed with, poked at, and even groomed the corpse. By contrast, Passion's attack on Gilka seemed to have been directed to one end only — the capture of her baby. And the carcass was consumed in the way that normal prey is consumed, slowly and with relish, each mouthful of meat chewed up with a few green leaves. We began to suspect that Gilka's first baby, little Gandalf, might have met a similar fate.

The following year Gilka gave birth to a healthy son, Orion. By this time she was terrified of Passion. They met, for the first time, when the baby was a few days old. Fortunately there were two adult males nearby. Passion approached to within ten yards, then stood staring at the tiny infant. Gilka instantly began to scream loudly, looking back and forth from Passion to the big males. As though they understood what was going on, the males charged over and, one after the other, attacked Passion. On that occasion it was she who fled, screaming.

During the next two weeks Gilka seldom travelled out of Kakombe Valley, where camp is situated. She seemed to be trying, desperately, to stay near the protection of the big males. I followed once when she set off from camp with Figan. For about ten minutes she managed to keep up, but gradually she fell further and further behind, handicapped by her physical disability along with the need to give frequent support to her newborn infant. Finally Figan disappeared along the trail ahead and Gilka was left on her own. I stayed with her. She nursed Orion, then sat for a while, staring down at her little son. Presently she began to feed. About two hours after she had lost Figan she heard the pant-hoots of Humphrey, calling from camp. Immediately she set off, back the way she had come, and joined him. They groomed for a while and then, when Humphrey left camp, Gilka trailed after him. Just as before, Gilka gradually got further and further behind and, after about twenty minutes, was once more left on her own.

It was inevitable that, sooner or later, Passion would encounter Gilka when there were no males nearby to help. It happened when Gilka, in the heat of midday, was resting with her infant in the shade. Orion was three weeks old. Pom arrived first, moving silently from the undergrowth. She stood watching mother and child for a moment, then lay down nearby. A more intelligent individual would probably have been instantly alert to danger. But Gilka, like Olly before her, was not characterized by any great intellectual prowess. She remained where she was, apparently not at all concerned. Five minutes later Passion appeared. Pom at once hurried towards her mother and reached to touch her back, a wide grin of excitement on her face. It was the sort of interaction that occurs between mother and daughter when they get close to a tree laden with delicious fruit. As one, Passion and Pom charged Gilka who, at first sight of Passion, had begun to flee. Gilka screamed and screamed as she ran, but there were no males nearby to respond to her desperate appeal for help.

Pom raced ahead of Gilka who veered to the side, trying to avoid her. At that moment Passion caught up, seized hold of Gilka and threw her to the ground. Gilka did not try to fight, but crouched protectively over her precious baby. Pom then flung herself into the fray, hitting and stamping on Gilka while Passion seized hold of the infant and bit at its head. Gilka vainly hit at her murderous attacker, while with her free hand she clung desperately to Orion. Passion bit Gilka's face and blood poured down from deep laceration on her brow. Then, working as a team, Passion and Pom together turned Gilka onto her back and, while the stronger Passion grappled with the mother, Pom seized the baby and ran off with him. Then she sat and bit deep into the front of his head. And so Orion was killed in the same brutal way as little Otta the year before.

Gilka wrenched herself free from Passion and raced after Pom but Passion was onto her in a flash, attacking her yet again, biting her hands and feet. Gilka, bleeding now from countless wounds, made a last valiant attempt to retrieve her mutilated infant, but it was hopeless. And then Passion, leaving Gilka, took the prey and hurried off, followed by Pom. Young Prof, who had watched the life and death struggle from the safety of a tree, climbed down and ran after his mother. Gilka limped after them for a short way but she was soon left far behind and after a few minutes she gave up and began to lick and dab at her wounds. The Passion family, meanwhile, vanished silently into the forest.

Probably we shall never know why Passion and Pom behaved in this gruesome manner. Gilka was not their only victim: Melissa lost one, possibly two infants to the killers, and, during the four-year period of their depredations a total of six other newborn infants vanished. I suspect that Passion and Pom were responsible for all these deaths. In fact, throughout that grim time only one female from the central part of the community range managed to raise her baby — Fifi. And then, after Passion became pregnant herself, the killings stopped. Not that she gave up immediately — we witnessed three further attempts but, for one reason or another, they failed. And then Pom also became pregnant and was no longer prepared to cooperate with her mother. After this the cannibalistic attacks came to an end and mothers, once again, could travel with their newborn infants without fear.

But for Gilka it was too late. She never really recovered from Passion's murderous attack. Although the lacerations on her hands seemed to heal, a few months later suppurating sores broke out on her fingers. And no sooner did they show signs of clearing up than they reappeared, worse than before. She had been lame before. Now she was truly crippled — sometimes she could barely hobble. She developed a chronic diarrhoea which never really left her, and she became increasingly emaciated. She was only fifteen years old, yet so poor was her physical condition that she never again resumed her periods of sexual swelling. Her reproductive days were over. She had been lonely before but she was infinitely more so now. Her closest companions at this time were two other childless females, the big, sterile Gigi and the immigrant Patti, who had not, as yet, given birth. But though we sometimes came upon these three peacefully fishing for termites together, or feeding on some seasonal crop of fruit, it was only when Gigi and Patti were visiting the home valley, for Gilka almost never moved further afield — she was too lame. When her friends set off for new pastures Gilka was left by herself.

She began to haunt our camp, more for companionship, I think, than the possibility of a hand-out of bananas. She would sit, a small lonely figure, gazing out over the valley, watching and waiting. Sometimes I sat close beside her, hoping that she would understand that I cared, that I wanted to help. Such was my relationship with her, such was her implicit trust in this human who had known and loved her since the carefree days of her infancy, that she even allowed me to smear antibiotic cream onto the terrible ulcers on her hands.

During these grim times Gilka's relationship with her elder brother acquired a new significance. True they were not often together, but when they were, Evered provided her with a very special kind of companionship. When he was nearby she became, for a while, relaxed and self-confident. Evered had been her solace once before, when old Olly died. She had been nine years old then, quite big enough to cope with life, but very much alone for she had no younger sibling, no close friends. And so, day after day, she had sought Evered's company. Often when she lagged behind, slow even in those days as a result of her polio, Evered had waited for her. And when, eventually, he had moved on and left her, she had sometimes seemed to trace his footsteps, following the same forest trails, stopping to feed where he had fed an hour or so before. Perhaps she had followed his scent, for chimpanzees can recognize individuals by their characteristic smell. Or perhaps she had glimpsed him, half a mile or so away, when both were feeding in the higher branches of tall trees.

As time went on Gilka and Evered had spent less time together, but always their relationship had remained friendly, characterized by long bouts of social grooming. Unlike other brothers, Evered was never seen to force his young sister to submit to his sexual interest during her periods of swelling. A few times he courted her, mildly shaking little branches, but when she ignored or avoided him, he left her alone. There had been many times when Gilka had clearly derived comfort from Evered's presence. After she had been threatened or attacked, for example, she had typically gone to sit close to Evered if he was in the same group. And then, quite visibly, she had relaxed. There was one occasion when Gilka and Fifi had an altercation in camp. We had set out a mineral lick and for a while the two females shared the block. But then Gilka accidentally bumped into Fifi who at once hit out at her. Gilka, enraged, hit back. In the face of such insubordination the higher-ranking Fifi attacked her childhood playmate. It was nothing serious, just a quick hitting and stamping, and Gilka, though she screamed and ran off a short distance, soon returned. She held out her hand, Fifi responded with a touch of reassurance, and both females resumed licking. Peace, I thought, had returned.

All at once, to my astonishment, Gilka gave a loud waa-bark of threat, and then, screaming, hurled herself at Fifi, hitting and grappling. Whatever was she up to? Then I understood: Evered had arrived. He stood surveying the battling females, hair slightly bristling. Suddenly Fifi too noticed Evered: quickly she retreated from the conflict, uttering little screams of fear — or was it fury! Gilka remained smugly at the salt, directing a few derisive barks at Fifi, and settled down to lick beside her big brother. After a suitable interval Fifi quietly approached the siblings, groomed Evered for a few moments, then joined in the licking — carefully keeping Evered between Gilka and herself! That was a good day for Gilka. And it must have been even more satisfying when, under the watchful eye of her big brother, she even dared to threaten Passion — with Evered looking on there was absolutely nothing that Passion could do!

There was one incident, towards the end of Gilka's short life, that illustrated vividly her inherent courage. The sound of loud baboon calls and the screams of a chimpanzee sent me scurrying through the forest. Eventually I came upon an incredible scene. Up in a small tree was a young adult male baboon, Sohrab by name, feeding on the carcass of a small bushbuck fawn. Close beside him, on the branch, was Gilka. To my amazement Gilka was trying to take some of his kill. Every time she reached for the meat, Sohrab turned and threatened her, showing his fearsome canines, raising his eyebrows so that the white lids flashed. When he did this Gilka screamed — but she did not move away. Instead, she tried again. Now Sohrab pushed at her with both hands, the meat in his mouth. And Gilka, weak as she was, fell from the branch. Fortunately she landed safely on another below, and after a few moments she climbed right back. When Sohrab again flashed his eyelids at her she screamed, louder than ever.

I watched, astounded. Below the tree many baboons were milling around in search of scraps, squabbling among themselves. At a discreet distance were two other female chimpanzees who seemed intimidated by the commotion and were just sitting, watching from their place of safety. But Gilka, weak and crippled, continued to harass the big male baboon. It occurred to me that she might have come upon the fawn herself, only to have it snatched from her by Sohrab. Surely only some sense of thwarted ownership could have led to such foolhardy behaviour.

Suddenly Gilka, screaming, raised both hands and slapped the baboon hard. Sohrab, infuriated, again seized the meat in his mouth and leapt at Gilka, grappling with her. This time both of them fell and landed together on the ground. Instantly one of the watching females raced over, grabbed the meat and pulled. Sohrab kept tight hold of one leg but the female managed to tear off the rest of the carcass and raced away with it. Many of the baboons and the other chimpanzee followed her. But Gilka climbed back into the tree after Sohrab. This, it seemed, was the last straw for him. Infuriated, robbed of the greater part of his prey, he leapt at this small audacious female, so that, once again, they both fell to the ground. And now he attacked her in earnest, pressing her to the ground and trying to bite her. Fortunately, though, he still had the meat in his mouth otherwise it would surely have gone badly for Gilka. As it was she was unharmed, though she screamed louder than ever, throwing a tantrum in her frustrated rage. All at once, Sohrab had had enough and ran off with the remains of the kill. There was no way that Gilka could keep up with him. She sat a while and stared where he had gone. And then she went to join the other chimpanzees, begging for a share. But they repulsed her with irritated threats and she soon gave up. Slowly she limped back to the scene of her conflict with Sohrab and searched the ground for any scraps left over from the feast. But the baboons had taken them all.

If only Evered had been nearby to hear her calls for help the incident would have had a very different ending. But he was far away, for it was the time when, after his defeat by Figan and Faben, he was forced to spend long weeks wandering in the north of the community range. Whenever he ventured back, he was invariably assaulted yet again by his two powerful adversaries. Then he would leave once more and stay away even longer. I had not realized, until then, that relationships between community males, individuals who had grown up together, could become so hostile — it seemed that the two brothers were actually trying to drive Evered from the community.

It was during these troubled times that I learned how the close, friendly relationship between Evered and his weakling sister sometimes benefited him as well as Gilka. One day, for example, I was in camp when Evered made one of his rare appearances. Perhaps it was no coincidence that, at the time, Figan and Faben were in the south of their range. But, even if he suspected that the brothers were not around, Evered was tense and nervous, glancing repeatedly from one side of the clearing to the other, startling at every rustle. Suddenly he stood, every hair on end, staring to the east where something was moving in the undergrowth. But it was only Gilka, and as she approached, uttering soft panting grunts of greeting, Evered relaxed. They groomed each other for a while, then left camp.

I followed them. For the rest of the day the two wandered about together, Evered, quite clearly, adjusting his pace to that of his sister. Several times he started off while she was still feeding, but after glancing back, lay down patiently until she had finished. Each time he got too far ahead during travel, he waited until she caught up. In Gilka's familiar, unthreatening presence I believe that Evered found the same kind of relaxation and comfort that he would have derived from being with his mother had she been alive. Surely this gave him added courage when, the following morning, he once more faced his bitter enemies.

But he was defeated yet again. Once more, he retreated to his refuge in the north and Gilka was left alone.

She was not quite twenty years old when she died. I saw her one day lying very still beside the swift flowing waters of the Kakombe Stream and I knew, even before I got close, that she would never move again. As I stood there I reflected on the long series of misfortunes that had dogged her, almost from the start. Her life, begun with such promise, had unfolded into a tale of infinite sadness. She had been an enchanting infant, filled with fun and an irrepressible gaiety despite the rather staid and asocial character of her mother. As a child she had delighted in male society, been intensely excited when, from time to time, Olly had joined a big group. Then, a born show-off, she would twirl and pirouette and somersault in an ecstasy of joy. And this was the chimpanzee who, her elf-like face transformed into that of some gargoyle, had become a pitiful cripple, and, of all the chimpanzees at Gombe, the most lonely.

It was dim and green in the forest, dappled with shifting brightness where the rays of the late afternoon sun fell through the rustling leaves of the canopy above. There was a murmuring of running water. And then, catching at the heart, the pure, hauntingly beautiful song of a robin chat. As I looked down on her, I knew a sudden sense of peace. Gilka, at last, had shed the body that had become nothing but a burden.

The brothers. Figan grooms Faben.

Curt Busse and David Riss during the fifty-day follow of Figan.

Figan displaying.
(Hugo van Lawick)

Grub and Maulidi.
(Hugo van Lawick)

One day, as Derek sat on the veranda of the house, the female baboon Algae suddenly climbed onto his shoulders. Derek, delighted, was transfixed.

Me with Goblin and, left to right, Msafiri and Hilali Matama. (Earl Bateman)

Gilka with Orion.

Because Humphrey is grooming her, Gilka is calm as infant-killer Passion (peeping from behind Gilka), Pom, and Prof peer at her third infant, Orion.

Passion and her family share a grisly feast, Gilka's three-week-old infant, Orion. (E. Tsolo)

Secure after being joined by Evered, Gilka threatens Fifi, who previously had attacked her during a squabble for possession of the mineral lick.

Flint interferes as Humphrey mates with Fifi. (P. McGinnis)

Charlie copulates with Fifi. Fifi is calm and relaxed. (P. McGinnis)

Charlie's courtship: he hits the ground with his knuckles. (P. McGinnis)

Kasakela males moving southward towards the Kahama community range. (C. Busse)

Kasakela patrol staring over hostile territory belonging to the Mitumba community to the north. (C. Busse)

Often males gang up to attack stranger females. Unlike attacks on community females, which are brief, attacks on strangers may last for more than ten minutes and result in serious wounding, even death. (B. Gray)

Brothers Hugh and Charlie, who initiated the split in the community. Displaying in parallel, they made a powerful team, terrorizing the northern males on many occasions. Charlie (closest to the camera) became alpha male of the Kahama community. (B. Gray)

9

Sex

<<<<<<<<<<<<<

OLLY'S LINE had seemed doomed to extinction despite the fact that she had left two independent offspring when she died. Her daughter, Gilka, had failed to raise a single child and for a while we had thought that her son, Evered, forced into exile, was doomed to wander on the outskirts of the community range, alone.

One Sunday morning Hamisi Mkono was walking along the lakeshore on his way to market. He was headed northward to the village of Mwamgongo that lies just outside the park boundary. He had, one by one, crossed each of the little streams that flow out into the lake from the watershed high up on the rift escarpment. First, after staff camp on the Kasakela Stream, comes Kasakela, then Linda, Rutanga and Busambo. Now he had reached the wide valley mouth where the Mitumba and Kavusindi Streams merge. There, up in an oil-nut palm not far from the beach, a chimp was feeding.

Curious, Hamisi moved a little closer, fully expecting that the chimp would run off — for this was the territory of the shy members of the Mitumba community, not yet used to humans. But the chimp calmly continued to feed — it was none other than Evered. A moment later Hamisi saw a second chimp peering at him from behind a palm frond — a female flaunting a fully swollen pink posterior. She didn't stay long for, despite Evered's calm acceptance of a human presence, she was nervous and soon climbed swiftly down and hurried off. Evered hastened to follow and the couple vanished into the thick forest of Mitumba Valley.

Here was no lonely exile! Not only was Evered in the company of a female, but a highly desirable female, at the height of her sexual receptivity. Even if he was being driven from his own community, he was making the most of the situation. Clearly he had persuaded one of the neighbouring females to accompany him on a consortship — an exclusive mating relationship. How many such sexual dalliances had Evered enjoyed during the months that he was driven from his community, we wondered?

It was about the time of that chance observation that Faben died, bringing an end to Evered's persecution, as without the support of his elder brother Figan's power diminished. And so Evered, although he remained submissive to the younger Figan for the rest of his life, was able to return and take up his position in the Kasakela community. This, however, did not bring his periodic romantic adventures to an end — rather they increased. For not only did he still, occasionally, consort Mitumba females, but he now found it easier to consort the females of his own community as well — adolescent females at the end of the infertile period, ready to conceive, and older females during the month when they resumed oestrus swellings between one child and the next. In addition, on the many occasions when pink females were not led on consortships but were surrounded by most or all of the males of their community, Evered could seize opportunities to mate with them along with the other Kasakela males. We suspect that Evered may have sired more infants than any other male of his time: Olly's genes, after all, will be well represented in the Gombe communities of tomorrow.

The goal of a consorting male is to keep his female away from rival males during the time when she is most likely to conceive — the last few days of her sexual swelling, before it suddenly becomes flabby and shrivels away. All the males at Gombe take females on consortships, but some do so more often or more successfully than others. Evered has shown consummate skill, not only in coercing females into following him, but in preventing their escape before he has had a chance to impregnate them. We were not able to record the progress of his dalliances with the shy Mitumba females, but his techniques have been carefully observed on countless occasions. A good example was the consortship that he initiated and maintained with Winkle in August 1978.

It began one morning when Evered came upon Winkle and her son Wilkie, then six years old, on the northern slopes of Kasakela Valley. As the big male approached, Wilkie ran up to greet him, jumping into his arms, then briefly grooming him. Winkle followed more sedately, with a few soft pant-grunts. She was just beginning a sexual swelling, and Evered was immediately interested, examining her posterior carefully, then sniffing his finger. When he had done, the two of them began a grooming session.

After ten minutes Evered moved away, then turned, and staring at Winkle, began, with quick jerky movements, to shake a leafy branch. Roughly translated this meant: 'Come! follow me!' (If the branch shaking is accompanied by a penile erection it means 'Come here! I want to copulate with you.') Winkle moved four steps towards Evered, then stopped. Evered shook the branch once more but in a half-hearted way, and when Winkle ignored him he did not press the point. After another ten minutes he tried again, and this time Winkle responded and she and Wilkie followed Evered as he set off, heading northward towards his favourite consort range.

After only a few minutes Wilkie, who had been last in line, climbed to feed on some choice fruits. Winkle, as though glad of the excuse, stopped immediately and sat to wait for her son. Evered turned and shook another branch but Winkle paid no attention. Over the next twenty minutes, Evered kept on repeating his summons and, as Winkle continued to ignore him, his shaking of the vegetation became more and more violent. It was obvious that his patience was gradually wearing thin, and finally it gave out altogether. With hair bristling, lips compressed, he leapt onto Winkle, pounding and dragging her until she pulled free and ran off screaming. Evered, panting from his exertions, once more summoned her, but she still refused to obey. She just sat looking at him, her screams gradually giving place to little squeaks, then whimpers.

Evered's patience was quite remarkable. He waited for almost thirty minutes, shaking branches from time to time in an irritated way. But as before he became increasingly frustrated and eventually disciplined her again, this time attacking her more severely. Now at last, when he stopped his pounding and summoned her to approach she responded instantly. Hastening to crouch before him, with nervous panting grunts, she pressed her mouth to his thigh, kissing him. And then, as is the way of male chimpanzees after aggression, Evered reassured her, grooming her until she relaxed under the gentle caress of his fingers. Once punishment has been handed out, then it is time to make amends, to restore social harmony. When, after twenty minutes, Evered again moved on, turned, and shook a branch, Winkle followed obediently, Wilkie, as before, bringing up the rear.

For some while they travelled thus, without further friction. On the ridge between Kasakela and Linda Valleys, they stopped to feed. An hour later Evered set off again, and in response to his now familiar summons, Winkle followed, but only a few steps at a time and with obvious reluctance. Quite clearly she was loath to leave her favourite haunts for the less familiar terrain to the north. Evered was more impatient now, and it was not long before he attacked her yet again. This was the worst of all: as he seized and pounded on her they tumbled together down a ravine, thudding from one large rock to another below, and then to a third. Winkle broke free and rushed away, screaming. But when Evered summoned her she quickly gathered up her son — who, scared by the conflict, was screaming loudly — and, carrying him on her back, she followed her implacable suitor.

For the next two hours Evered led the way relentlessly further and further northward. Three more times he attacked Winkle, once when she baulked at crossing Linda Stream, once when she suddenly ran to the south, startled by the sudden shouting of fishermen from the beach nearby, and finally when she made her last attempt to resist him just before they moved down into Rutanga Valley.

Not until it was nearly dark did the little group settle down for the night. Wilkie shared his mother's nest as usual, and surely the contact with his small, familiar body gave her some comfort after the bruisings and batterings of the long day.

The next morning things were very different. Winkle, now that she had finally moved into unfamiliar territory, was only too anxious to stay near Evered and, for the most part, followed him readily whenever he moved on. His episodes of branch shaking became ever less frequent and less vigorous. By 10.30 they had already reached Kavusindi, and that night they slept together in Mitumba Valley, near the beach, where Evered almost always takes his females. And there, for the next eight days, they would remain.

Once they had settled down, safe from discovery by other Kasakela males, Evered became benign and tolerant. If, when he was ready to leave, Winkle was still feeding, resting, or grooming her infant, then he stretched out on the ground and waited patiently. He groomed her often, and during the heat of midday all three often lay close together on the ground. Evered was very tolerant of Wilkie, too, sometimes grooming him for a while and even, on a number of occasions, sharing food when the infant begged. But for the most part Wilkie was sulky and depressed, for he was going through the final stages of weaning. He spent much time sitting in contact with Winkle, and, desperate for reassurance as a result of the drying up of her milk, constantly demanded her attention.

Winkle was fully pink from the third day of the consortship onwards. She was fertile and, towards the end, at her most sexually attractive and receptive. Yet Evered mated with her but seldom — never more than five times in one day. When he did court her, Winkle responded quickly and calmly. It was all so peaceful, like some idyllic honeymoon.

It is not only Evered who becomes benign and tolerant once he has led his female to his chosen consort range: it is the rule among the Gombe males. The aggressive bullying ceases once the male has achieved his goal, and he is then prepared to adjust his daily routine to that of his lady. I remember once when Figan took Athena northward to Rutanga Stream. She was extraordinarily reluctant to accompany him and it was a harrowing day for both of them. Eventually, however, by dint of repeated violent displays — and without any fighting — Figan had his way. The following morning Athena clearly wanted a lie-in. Figan rose at the usual time and went to sit below Athena's nest. She peered down at him, gave a soft grunt, a sleepy 'Good morning', and stayed right where she was. After ten minutes Figan, gazing up, shook a little clump of vegetation. No response from above. Eight minutes later he tried again, but she continued to lie in bed, paying Figan no attention. Even when he performed a swaggering display she still ignored him. And so, eventually, he moved off without her to attend to his own pressing need for breakfast. The tree into which he climbed, laden with succulent *mmanda* figs, was not far away, but even from the topmost branches he could not see Athena. After stuffing food into his mouth for a few minutes he hastened down through the branches, ran back a short way, peered anxiously towards her nest and then, assured that she was still there, returned to the fig tree. During the next forty-five minutes he interrupted his meal five more times in order to ascertain that Athena had not escaped. By the following day Figan had led Athena much further to the north. Then he could relax and the remaining thirteen days of their consortship were peaceful and calm.

How different is the situation that prevails when a sexually attractive female is surrounded by a whole group of adult males. Then, if she is a popular partner, tension builds up as her suitors vie with one another for the opportunity to mate. Under these conditions the female may copulate with six or more males in ten minutes or so. And whenever there is some excitement in the group, such as reunion with other chimps or arrival at a food source, this typically triggers a renewed outburst of sexual activity. Old Flo, during her heyday, was once mated fifty times in one twelve-hour period. And only too often, with tension running high, fighting breaks out, sometimes for the most trivial of reasons. Even though the female herself is seldom the victim, the situation clearly subjects her to a certain amount of stress.

It may well be that the calm and friendly atmosphere of the consortship is more conducive to conception. Certainly eight months after Winkle returned from her honeymoon with Evered she gave birth to a daughter — we called her Wunda (it makes a better name spelled that way) since, for the first time in Gombe's history, humans observed the birth. And as eight months is the gestation period in chimpanzees, Wunda, without a shadow of doubt, is Evered's daughter.

When a female chimpanzee becomes pregnant her condition seems to remain a secret, at least for a while. There is no signal comparable to the sudden change in the colour of the posterior of the pregnant

female baboon. There does not seem to be any special odour, or pher-
omone, to advertise her condition to the males. Moreover, for the first
few months of her pregnancy she is likely to develop sexual swellings
as usual, and at least the first of these may arouse the interest of the
adult males. This leads to some absurd situations when males wear
themselves to the bone in order to lead away reluctant females who
are already impregnated with the seed of rivals.

Often a male has to work very hard indeed to take and keep a female
on a consortship. If his lady conceives, this effort will have been more
than worth it. But, of course, he has no way of knowing. Probably
this is why some males go to so much trouble to take their females on
two consortships in succession, because this, in a way, safeguards the
investment he has already made. If he failed to impregnate the female
the first time around, the second honeymoon will give him another
chance. And it will prevent her from going off with a rival. Even if she
is already carrying his child, it may still be worth his while since he
will effectively ensure that she is not subject to the stresses and strains
of an excitable sexual gathering, a situation which could have a dam-
aging effect on her — and his — unborn child. Evered sometimes took
his females on three successive honeymoons of this sort.

Each adult male has his own particular style when it comes to consorting. Evered goes in for lengthy consortships — many of his were
considerably longer than the ten days he spent with Winkle. Once he
is thought to have wandered in the north with one of the Kasakela
females for almost three months, although we cannot be sure they
were together all the time.

Other males go in for very short consortships. They try to initiate
the relationship not during the early stages of swelling but when the
female is already fully pink. There are distinct advantages for the male
who can carry this off. For one thing the female, being highly sexually
receptive, is more likely to cooperate with him. For another, he does
not have to maintain the relationship for so long and this is important
if he is working to keep up his position in the hierarchy: the longer
his absence, the more likely it is that, on his return, he will have to
face challenges from one or more of his rivals.

But the strategy has its drawbacks too. It is not easy to elope with a female who is at the height of her attractiveness. Indeed, if she is sexually popular, it may be impossible since she will be surrounded by a number of adult males watching every move she makes. The would-be consort male must keep close to her and be prepared to seize any opportunity to try to lead her away. Of course, even if he fails, his constant proximity to her will give him maximum opportunity to copulate with her and that might make all the difference to his chances of fathering a child.

One of the chief proponents of the short-and-sweet consortship was Satan. His technique was interesting. He not only maintained close proximity to the female with whom he wished to make off, but he groomed her frequently as well. And then, having thus demonstrated his kindly nature — 'See what a charming partner I shall be' — he waited his chance. If, for any reason, he and the female were temporarily separated from the others Satan quickly shook vegetation, led off in the opposite direction from the group, and hoped she would follow. A couple of times, when the female stayed up late in the evening, feeding ravenously to make up for a scanty intake during her sexually busy day, Satan stayed up too. And then, when her meal was over and the other males were safely in their nests, he tried to lead her a short distance away. If this was successful he would then rise very early the next morning, and, after rousing the lady, suggest to her that they make a hasty get-away.

Ploys of this sort work only if the female cooperates. If she refuses to follow and the male attacks, her screams are sure to bring one or more of her other suitors racing to the scene. Satan did well in this respect and often succeeded in setting off with quite popular females. But this did him little good since almost always the female, after staying with him for a couple of days, then gave him the slip and reappeared, still fully pink, in the central part of the range. There the other males made haste to copulate with her, making up for lost time. Yet despite the obvious failure of his strategy Satan continued trying.

Some males, using a technique that is the exact opposite of the short-and-sweet method, set off on consortships with females who are completely 'flat' — who show no signs at all of developing a sexual swelling. Sometimes they even consort females who have recently deflated, who may have just returned from an extended consortship with another male. It is one way for a low-ranking male to get a female to himself, for his superiors will not be interested in her at this stage and will not object to his manoeuvre. If he succeeds in taking her off, succeeds in keeping her with him until she does become fertile, he will be in clover. He will know a halcyon few days when he has a female, at the peak of her swelling, all to himself. He can mate her whenever he likes, without fear of interruption by his superiors. Moreover, unless she is already pregnant he will have a good chance, in this peaceful setting, of siring a child, propagating his genes — which, after all, is what sex is all about.

The main problem for a male trying to lead away a flat female is that during this 'cold' phase of her sexual cycle she is usually particularly reluctant to accompany him. We observed the entire course of what may well have been young Freud's very first attempted consortship. He was fifteen years old at the time and his chosen partner was Melissa's daughter Gremlin. She was completely flat. She had only recently returned from a week with Satan. And she most emphatically did not want to go anywhere with Freud.

When I met them Gremlin was sitting by a tree trunk, and Freud was glaring at her, shaking branches. Only after he had displayed around her several times, violently swaying the vegetation, did she eventually trail after him, heading north. She kept looking back, her lips pouted, and every so often I could hear a soft whimper of distress. Quite clearly she wanted to rejoin her mother, with whom she had been travelling earlier in the day. But whenever she turned and tried to move back the way they had come Freud shook branches at her. If she refused to follow he stood upright, shaking and swaying the vegetation in yet another magnificent display. Gremlin pushed her luck to the limit, ignoring him until it seemed that an attack was inevitable. But then, at the last minute she hastened towards him with panting grunts and gestures of appeasement. This was usually followed by a brief grooming session, after which Freud tried again. He was two years younger than Gremlin but already much stronger, and in a fight she might well have been hurt. And so eventually she gave in.

However, she soon worked out a way of coping, her own unique form of protest. After travelling a few steps in the required direction she climbed a tree and began to feed. Freud, after looking up and half-heartedly shaking a little tuft of grass, settled down to wait. He waited and waited and waited. He lay down and closed his eyes. He sat up and groomed himself. Finally, after almost an hour had passed, he began to show signs of mounting impatience, scratching himself ever more vigorously while his glances towards Gremlin became more and more frequent. Eventually he performed another series of spectacular displays below her — and even then she just sat without moving, watching him. Only when Freud actually leapt, bristling, into her very tree, did she at last capitulate, climbing down and reaching to touch him appeasingly.

When he moved off, still heading north, Gremlin followed. But after a few yards, she climbed another tree and again began to feed! Never have I seen a chimpanzee climb so many trees in such a short space of time. Anything as an excuse to delay. And each time Freud waited as before, grooming himself or sprawled out on the ground, until she condescended to follow him again — for another couple of yards. After five hours they had travelled only some five hundred yards! When, about one and a half hours before the usual bedtime, she climbed yet another tree and constructed a leafy nest, Freud, after gazing up, gave an audible sigh, then resignedly made his own nest nearby.

They were still well within the central part of the community range when, the next day, they encountered a couple of other Kasakela males. This marked the end of Freud's attempted consortship, and Gremlin was able to rejoin her mother.

It is quite clear that a female prefers some males to others; equally there are certain individuals whom she may actively try to avoid. The aggressive Humphrey was, understandably, feared by many of the females. But, although a female can sometimes bring an unwelcome relationship to an end — by calling out and attracting other males or by seizing an opportunity to escape — for the most part she must submit to the whims of any male who desires to take her away. And while there are times when a female appears to follow a male willingly, this may simply be the result of bitter punishment for disobedience on previous occasions.

Once when Passion, quarter swollen, refused to follow Evered to the north, he attacked her four times, very severely indeed, in less than two hours. During the third of these assaults, Passion's hand was badly injured and afterwards she could not put it to the ground. Thus lamed she was even slower to obey Evered's imperious demands and his fourth attack was the worst of all. This time her frenzied screaming, along with the calls of her agitated offspring, Pom and Prof, attracted the attention of two males. When they arrived, hair bristling, to see what was afoot, Evered hastened to greet them and then, without as much as a backward glance at Passion, went off with his two friends. Passion, who was still uttering little whimpers and no doubt feeling very sorry for herself, must have been delighted to see him go.

But she was not to be rid of him so easily. The next day he found her again, and this time she hastened to obey his imperious summons at once, limping after him as quickly as she could. She had learned her lesson well. Evered, as far as we know, kept her away from the other males for nearly two months — throughout two periods of full swelling. When she finally reappeared in her usual haunts, she was pregnant — presumably with Evered's child.

One interesting aspect of Evered's long consortships is the fact that he quite often copulates with his females when they are not fully pink. This is very unusual behaviour in the wild. An adult male almost never courts a female except during the ten days of her maximum swelling, and she, for her part, does not willingly respond if he tries to force his attentions on her at any other time. If he persists she typically becomes fearful and tries to avoid him. But Evered, during his long consortships with two females, Athena and Dove, copulated with them on a number of occasions when they were either completely flat or, at most, one-quarter swollen. And each time they accepted his sexual advances quite calmly. Probably it was the same when he spent weeks with other females, but we were not there to see it.

This whole set-up — the prolonged period of the exclusive relationship, the calm and relaxed atmosphere that prevails, and the unusual sexual interactions — suggests that chimpanzees have a latent capacity for the development of more permanent heterosexual pair bonding: a relationship more similar to the pattern of monogamy —

or at least serial monogamy — that has become the cultural tradition in much of the western world.

Even during the most seemingly idyllic of consortships, however, the seeds of unfaithfulness are present. Once Evered was with Dove, in his favourite consort range in the north, for almost two months. It was on a bright sunny morning, towards the end of this period, that his loyalty was challenged and found wanting. For half an hour after leaving their nests Evered and Dove, along with Dove's juvenile daughter, had fed on pale yellow blossoms. And now the two adults sat close together and groomed each other while the child played by herself in Evered's empty nest nearby. Dove, at the time, was flat and, as we subsequently discovered, pregnant with Evered's child.

Suddenly there was a rustling in the undergrowth nearby. Evered tensed and stared towards the sounds, his hair bristling. Only a few days before his little group had fled silently southward when they heard the pant-hoots of Mitumba community males nearby, and clearly Evered was prepared to lead another retreat now. As a chimpanzee began to climb a tree about a hundred yards away, Evered showed his teeth in a silent grin and, as a second chimp followed the first, he reached to touch Dove, seeking reassurance.

But after a few moments he relaxed, recognizing two members of his own community — Sherry, a young male in his prime, and Winkle, full swollen. Another honeymoon couple! Evered stared for a few moments and then, his hair still bristling, rushed towards them, climbed swiftly into their tree, and began to shake branches at Winkle. Whether or not she had any inclination to obey his summons we shall never know for Sherry, normally subordinate to Evered, was instantly prepared to defend his rights. He charged at Evered and attacked him. The fight did not last long and soon Evered, smaller and lighter than Sherry, retreated, screaming. But he did not leave and after a few minutes Sherry attacked again. This time Evered was actually kicked from the tree and fell some distance to the ground.

Still screaming and thoroughly routed, he returned to his Dove. She had remained where he left her, watching the whole performance. As he sat beside her, whimpering, and licking a bleeding toe, she began to groom him and gradually he quietened. But he continued to gaze towards Winkle until, following Sherry, she and her provocative swelling vanished into the forest.

That incident highlights the powerful effect of the female's swelling in arousing the sexual desires of a male. It was not clear whether Evered was merely wanting to steal a quick copulation with Winkle, or whether, as I suspect, he was attempting to terminate her relationship with Sherry and take her for himself. If he had been successful in such a manoeuvre, what would have happened to Dove? Would Evered, like the old male Leakey a decade earlier, have tried to keep both females with him? This seems unlikely. Almost certainly, Dove, flat and relatively uninteresting, would have been abandoned in favour of the pink and alluring Winkle.

Dove would then have been in a highly vulnerable position. She would have been left without male protection in an area relatively unfamiliar to her, for her favourite haunts are in the south. And there she and her child would have been at the mercy of the males of the powerful Mitumba community.

10

War

T HE KASAKELA PATROL moved forward slowly and cautiously as it penetrated ever deeper into the territory of the Mitumba community. Satan was in the lead; five other males and Gigi, fully pink, followed close behind. All had their hair erect, bristling with apprehension and excitement. First one and then another bent to sniff the ground. Evered picked up a leaf and smelled it carefully; Figan stood upright to sniff the lowest branches of a tree. Repeatedly they paused to listen, staring into the dense undergrowth on either side. It was a wind-still day and the forest was silent but for the periodic shrilling chorus of the cicadas. Suddenly a twig snapped, a sharp, brittle sound. Satan turned to the others, his face split by a wide grin, part fear, part excitement — a gash of white teeth set in bright pink gums. Silently he embraced Jomeo who was behind him. Figan and Evered also threw their arms around one another, Mustard reached to touch Goblin. Like Satan, all were grinning hugely.

As they stood there, silently staring towards the source of the sound, another twig snapped. Leaves crunched under a heavy tread. And then the chimps relaxed as the large dark shape of a bushpig appeared, rootling his way through the undergrowth. Intent on his own concerns he never even noticed his audience, and he soon disappeared.

Satan moved on again, but when he looked back and saw the others were not following he paused: he was not prepared to go on by himself. After a moment, however, Jomeo followed, then the rest.

Ten minutes later the soft whimper of an infant was heard just ahead. Instantly, after glancing at each other, the males and Gigi raced towards the sound. Just as they reached a tall, sparsely foliaged tree a female leapt down. She might have got away, but her infant, between two and three years old, was still up in the branches, screaming now in fear. The mother raced back, seized her child and once more hurled herself to the ground. But she had lost valuable time — the Kasakela patrol was upon her. Goblin was the first to seize the stranger, hitting and biting her and stamping on her back. A juvenile, who had also been in the tree, quietly climbed down and vanished into the dense bush. Satan and Mustard leapt to join Goblin as he continued the attack, and a moment later Figan and Jomeo hurled themselves into the fray.

During this fierce assault, Evered seized the infant and charged off through the bushes, flailing it against the ground as though it were the branch of a tree. Then, hurling the little body ahead of him, he turned back, racing to join the other males who were still attacking the mother. Gigi was there too, on the outskirts of the mass of screaming, yelling bodies, getting in a hit whenever opportunity offered.

Some ten minutes after the start of the attack the female managed to pull herself free and, still screaming in terror, clambered into a tree. Goblin was the only male who followed. He attacked her briefly, then watched as Gigi, obviously determined to have the last word, climbed up and delivered a final series of hits. The stranger pulled free, took a huge leap to a neighbouring tree, another to the ground, then headed towards her infant, who was still screaming in the undergrowth. The whole encounter had lasted some fifteen minutes. There was a great deal of blood on the flattened vegetation where the worst of the fighting had taken place, and some under the tree where Goblin and Gigi had meted out the final punishment.

For the next five minutes the Kasakela chimpanzees, in a state of excitement that bordered on frenzy, charged back and forth around the scene of conflict, dragging and hurling branches, throwing rocks, uttering the deep, low-pitched hooting calls that sound like roaring. Eventually, still in a noisy and boisterous mood, they turned and moved back the way they had come.

At least once a week the Gombe males, usually in groups of not less than three, visit the peripheral areas of their community range. There is no clearly marked boundary between neighbouring social groups; usually, in fact, there is an area of quite extensive overlap between them. When the males discover some good source of food in such an overlap zone, they often go back the next day to feed, accompanied then by females and youngsters. On expeditions of this sort, the chimpanzees typically ascertain the whereabouts of their neighbours before commencing to feast. Thus when they reach some high ridge overlooking neighbouring territory the expedition members pause and very carefully scan the country ahead. If all seems clear they usually utter loud pant-hoots, then listen intently. Only if they hear nothing, or if there is a reply from very far away, will they advance confidently and begin to feed.

Sometimes, as a group of chimpanzees wanders along foraging, pausing occasionally to rest and groom, the adult males suddenly begin to travel in a brisk manner, heading towards some outlying part of their community range. This sudden sense of purpose, this air of determination, usually indicates that they are setting off to monitor the whereabouts of their neighbours. At this point mothers and young who have been travelling with the males typically drop back — except pink females, who usually tag along behind.

When patrolling males detect the presence of strangers, they begin to move cautiously, sniffing the vegetation, alert to the slightest sound. The discovery of discarded wadges of fruit peelings or abandoned termite fishing tools arouses immediate interest. If a fresh sleeping nest is seen, the males usually climb up to investigate it thoroughly, then display wildly through the branches until it has been virtually destroyed. If they actually see chimpanzees from the neighbouring community their response will depend on the size of that group relative to theirs — particularly with regard to the number of adult males. If one of the groups is larger than the other, or has more adult males, then the smaller one typically retreats discreetly and silently to a safer place. If the other males notice they will call loudly and give chase but, provided there is more than one male in the fleeing group, the pursuers will not try to catch up: they are content simply to provide a show of strength. If honours are about equal — with similar numbers of males in each patrolling group — then members of both sides, usually keeping several hundred yards apart, hurl threats at each other. First one group, and then the other, performs wild displays, charging through the undergrowth, slapping and stamping on the ground, drumming on tree trunks, throwing rocks, and all the while uttering loud, fierce calls. Finally, after half an hour or more, each side retreats towards the safe central part of its own home range. This vigorous and noisy behaviour serves to proclaim the presence of the legitimate territory owners and to intimidate the neighbours. Fighting is not necessary.

It is when two or more males encounter a lone stranger, or a couple of stranger females with infants, that fierce and brutal attacks take place. Indeed, if patrolling males hear the calls of an infant in some outlying part of their range and suspect the presence of a mother from a neighbouring community, they sometimes stalk her, persisting for an hour or more in their attempt to hunt her down. And, if they are successful, they will attack. A stranger male may be attacked also, but during the course of our years of research at Gombe we have observed only two, relatively mild, attacks on males from neighbouring communities, as compared to eighteen severe assaults on stranger females. Males, after all, are far more dangerous adversaries, particularly when they are strangers and their strengths and weaknesses unknown. Of course a lone male could be defeated by a group of males — but he might inflict some serious injuries on one or more of his aggressors during the battle. A female, especially if she is protecting an infant, poses no danger to her assailants.

Why are these females attacked so savagely? In some mammalian societies — those of lions and langur monkeys, for example — a male who has defeated the leader of a group and taken over the females will sometimes kill all small infants. With luck, his newly acquired females will become sexually receptive sooner than would have been the case had they weaned their infants normally. The new leader will then have a double advantage: first, he will be the father of all subsequent babies born into his group; second, he has eliminated some of the offspring of his defeated rival who, had they survived, would have competed with his own. In terms of evolutionary theory, this exercise will be to the reproductive advantage of the killer male if it leads to a higher proportion of his biological kin in future populations than would otherwise have been the case.

The attacks observed at Gombe, however, were very clearly directed at the adult females themselves. Although there were four occasions when infants were, in fact, killed, each time it seemed that this was incidental to the savage assaults on their mothers. Whenever it was possible to get a good view of the victims after they had escaped we could see that they had been badly wounded, whereas their infants, except for the unfortunate four, seemed to be unharmed. It would be relatively easy for even one male on his own to seize an infant from its mother, and kill it, if that was his goal. It seems, then, that the attacks are an expression of the hatred that is roused in the chimpanzees of one community by the sight of a member of another. Strangers of either sex may trigger this hostility, but the unthreatening females are attacked far more often. In this way males dissuade them from moving into their territory — if, indeed, they survive — and food resources within the community range are protected for their own females and young.

There are, however, certain times when females are safe from savage intercommunity aggression of this sort. Late adolescent females typically move into neighbouring communities during periods of oestrus. And not only are they tolerated by the adult males there but, when fully pink, they may be actively recruited by patrolling males — who clearly find them highly sexually stimulating. Sometimes a young female remains in the new community after becoming pregnant. This is a tough decision. For one thing, her presence will be intensely resented, at least to start with, by the resident females. For another, she is effectively severing all ties with her family and the companions of her childhood since, once she has given birth, she will not be able to return to her own community. If she tried she would run the risk of being brutally attacked — unless, again, she were fully pink. We have observed a few encounters between community males and stranger females in oestrus and, although there were some attacks, there were many copulations also. But such incidents are rare — most females are carefully guarded by their own males when pink.

There is absolutely no question but that intercommunity encounters are highly attractive to some of the males, particularly when they are between fourteen and eighteen years old. Once I followed as Figan, Satan and young Sherry travelled slowly along the southernmost ridge of Mkenke Valley, at that time part of the overlap zone with the powerful Kalande community to the south. Suddenly Figan stopped, hair on end, and staring southward gave a loud call of alarm. I followed the direction of his gaze and there saw a group of at least seven adult chimpanzees. Obviously they were members of the Kalande community and now, alerted by Figan's call, they began to display, vigorously and noisily.

The three Kasakela males ran silently northward for a short distance, then stopped and looked back. As the strangers displayed again, moving in our direction, Figan and Satan turned and fled in silence, back to safety. But Sherry, with adolescence just behind him, did not follow at once. He stood watching the advancing strangers, absorbed and fascinated. Only when two adult males charged to within fifty metres did he turn and run after his companions. And later in the day he left Figan and Satan and returned, by himself, to the Mkenke Valley ridge. There he climbed a tall tree and sat, staring southward, for over half an hour. It was as though he simply had to have just one more look.

Another young Kasakela male, Sniff, once taunted a large group of Kalande chimpanzees, including at least three fully adult males, quite by himself — his two companions had fled. The Kalande group was in a shallow, steep-sided ravine, calling loudly and charging about in the undergrowth. Sniff, uttering deep roar-like hoots, performed a spectacular display along a trail near the top of the ravine. As he charged he hurled at least thirteen huge rocks down onto the strangers. An occasional missile — a stone or a stick — flew up from the undergrowth below, but they fell far short of Sniff. Only when two Kalande males raced towards him did Sniff retreat. And he was still roaring his defiance, still slapping and stamping on the ground and drumming on the tree trunks, when he caught up with his chicken-hearted companions.

1974 marked the start of 'the four-year war' at Gombe. Ten years after I arrived at Gombe, the community whose members I had come to know so well began to divide. At that time, towards the end of Mike's reign as alpha, there were fourteen fully adult males: six of them, including the brothers Hugh and Charlie and my old friend

Goliath, began to spend more and more time in the southern part of the community range. Sniff, who was an adolescent at the time, and three adult females with their young, also became part of what we called the 'southern sub-group'. The 'northern sub-group' was much larger, with eight adult males, twelve females and their young.

As the months passed, the relationship between the males of the two sub-groups became increasingly hostile. The northerners tended to keep out of the area used by the breakaway group, but every so often, led by Hugh and Charlie, the southern males moved north-wards. And, because they almost always made such forays in a tight-knit group, and because of the fearless natures of Hugh and Charlie, the northern males usually avoided them. Still, though, the two oldest of the northern males, Mike and Rodolf, sometimes wandered about peacefully with Goliath, the oldest southerner.

Two years after the first signs of a split, it was clear that the chim-panzees had now become two distinct communities, each with its own separate territory. The southern 'Kahama community' had given up the northern part of the area where it had once ranged, while the 'Kasakela community' found itself excluded from places in the south where it had previously roamed at will. When males of the two com-munities encountered one another in the overlap zone between the two, they typically hurled noisy insults at each other, displayed long and vigorously, then retreated, each side back into the safe heartland of its own newly demarcated territory. But even then the three oldest males sometimes renewed their friendship.

For a year things continued in this vein. And then came the first brutal attack by Kasakela males on a Kahama male. It was observed by Hilali and one of the other field staff. The assault began when a Kasakela patrol of six adult males suddenly came upon the young male, Godi, feeding in a tree. So silently had the aggressors approached that Godi was not aware of them until they were almost upon him. And then it was too late. He leapt down and fled, but Humphrey, Figan and the heavyweight Jomeo were close behind, running shoulder to shoulder, with the others racing after them. Humphrey was the first to grab Godi, seizing one of his legs and throwing him to the ground. Figan, Jomeo, Sherry and Evered pounded and stamped on their vic-

tim, while Humphrey pinned him to the ground, sitting on his head
and holding his legs with both hands. Godi had no chance to escape,
no chance to defend himself. Rodolf, the oldest of the Kasakela males,
hit and bit at the hapless victim whenever he saw an opening and Gigi,
who was also present, charged back and forth around the melee. All
the chimpanzees were screaming loudly, Godi in terror and pain, the
aggressors in a state of enraged frenzy.

After ten minutes Humphrey let go of Godi. The others stopped
their attack and left in a noisy, boisterous group. Godi remained motionless for a few moments, lying as his assailants had left him, then
slowly got to his feet and, giving weak screams, stood gazing after
them. He was badly wounded, with great gashes on his face, one leg
and the right side of his chest, and he must have been badly bruised
by the tremendous pummelling to which he had been subjected. Undoubtedly he died of his injuries, for he was never seen again by the
field staff and students working in the Kahama community range.

Over the next four years, four more assaults of this sort were witnessed. The second victim was the young male Dé. He was equally
badly wounded as a result of a twenty-minute battering inflicted by
Jomeo, Sherry and Evered. Again Gigi was present, and that time she
actually joined the males in their attack. Dé, emaciated and with a
number of unhealed wounds, was seen for the last time one month
after the attack. Then he too disappeared for ever.

The third victim was, for me, the most tragic of all. It was none
other than my old friend Goliath, the second chimpanzee who had
ever allowed me to approach him closely. Goliath who had been top-
ranking before Mike's reign, was always one of the boldest and bravest
of the adult males. Why he had moved to the south at the time of the
community division will always remain a mystery to me. The other
Kahama males had, from the start, shown close affiliations with each
other and spent much time together. But Goliath had always seemed
to be more friendly with the Kasakela males who, at the end, so suddenly and brutally attacked him. He was old and frail when it happened, with his once powerful body wizened, his once glossy black
hair faded and brown, his teeth worn down to broken stumps.

One of the students, Emilie, was present during the attack that led to Goliath's death. What shocked her most was the terrifying rage and hostility of the five aggressors — Figan and Faben, Humphrey, Satan and Jomeo.

'They were definitely trying to kill him,' she told us afterwards. 'Faben even twisted his leg round and round — as though he was trying to dismember an adult colobus after a hunt.'

When the assault was over Emilie followed the assailants back to the north and recorded their wild excitement. Repeatedly they drummed on tree trunks, hurled rocks, dragged and threw branches. And all the time they called out, as though in triumph.

Goliath, like the other victims, had been horribly wounded. He managed to sit, but with difficulty, and as he gazed after his one-time companions he was trembling violently. He cradled one wrist with his other hand as though it was broken, and his body was covered with wounds. The next day we all turned out to search for him but he too vanished without trace.

After the death of Goliath, only three Kahama males remained — Charlie; Sniff, now a young adult male; and Willy Wally, still crippled as a result of the 1966 polio epidemic. Hugh had vanished, probably killed like the others.

Charlie was the next to go. No one saw the attack on him, but fishermen reported hearing the sounds of fierce conflict and, after searching in the area for three days, field staff found Charlie's dead body lying near the Kahama Stream. The nature of his terrible injuries was proof enough that he too had been killed by the Kasakela males.

It was clear by then that the Kahama males were doomed: sooner or later the remaining two would be hunted down and killed. But I was deeply shocked when the next victim was neither of them but one of the three females, Madam Bee. I suppose I should have been prepared for this, knowing of the brutal attacks on stranger females. But Madam Bee was not a stranger, and I had thought that the Kasakela males, once they had disposed of their Kahama rivals, would probably try to take back the three females who had 'defected' to enemy ranks.

Like Goliath, Madam Bee was old. And she was even more frail, with one arm paralysed by polio. At the time of the fatal assault she had already been subjected to a series of attacks and was weak from a number of unhealed wounds. Yet this defenceless female was set upon in the same vicious way, pounded and hit, kicked and dragged and rolled over. After the final battering she lay face down, completely motionless, as though she were dead. But, as the aggressors displayed away, calling noisily, she somehow managed to drag herself into some thick vegetation.

She hid so well that it took two days of diligent searching to find her — and then it was only because her adolescent daughter, Honey Bee, was seen feeding in a tree above. For the next two days the stricken female lay on the ground, sometimes dragging herself a few feet only to collapse again. Gradually she became weaker and was seized by uncontrollable spasms of shivering. Four days after the attack she died.

There was nothing we could have done to prevent her death. If she had recovered, there would have been no future for her: even healthy males in the prime of life had not been able to avoid the implacable hostility of their Kasakela enemies. We did try to ease her passing by taking food and water to her where she lay, but she accepted very little. Only in the presence of her adolescent daughter did she appear to find some comfort, for Honey Bee remained close by throughout those last cruel days, grooming her mother and trying to keep the flies from her wounds.

Willy Wally was the next to vanish. And then, for a year, Sniff was the lone survivor of the Kahama males, confined to a tiny area sandwiched between the Kasakela community to the north and the powerful Kalande community further to the south. I hoped so desperately that, against all odds, Sniff might make it. If only he could somehow gain admittance to the Kalande ranks. Or move to some unclaimed land outside the park boundary, east of the rift. He was so young, and so well-loved.

I remember when, in 1964, Sniff's mother had visited camp for the first time. While she hovered nervously in the bushes at the edge of the clearing, Sniff, with his insatiable curiosity, approached my tent, lifted the flap and peered inside. He had not seemed at all scared when he saw me, peering out! We had watched him grow up, from an engaging and playful youngster to a sturdy adolescent. We had been deeply touched when, after the death of his mother, Sniff (then eight years old) had adopted his fourteen-month-old sister. Still dependent on her mother's milk, she had only survived for three weeks, but during that time he had carried her everywhere he went, sharing his food and his nest at night, doing his best to protect her during the frequent aggressive incidents that broke out in camp at that time of intensive banana feeding.

But Sniff was brutally murdered like the others. Hunted down, attacked and left incapacitated, bleeding from innumerable wounds and with a broken leg. Once again we all went out to search for him: once again we failed to locate the place where he had crept away to die. His passing marked the end of the Kahama community. For a while there were occasional glimpses of the two remaining adult females and their infants, but then they too vanished. Probably they met the same fate as the rest of that ill-fated little group. Only the adolescent females had been, from the start, immune from the violence.

The four years from early 1974, when Godi was attacked, until late 1977, when Sniff was killed, were the darkest years in Gombe's history. Not only was an entire community annihilated but, in addition, there were the cannibalistic attacks of Passion and Pom, the gruesome feasting on the flesh of newborn babies. And it was during that same time that the Zairean rebels had landed on Gombe's sandy beach and plunged us into the nightmare weeks that followed. I suppose we should thank God that the human drama, though it resulted in untold mental anguish, at least claimed no lives.

The kidnapping, despite the shock and misery, did little to change my view of human nature. History is peppered with accounts of kidnap and ransom and there have been many studies, particularly in recent years, of the effect that incidents of this sort may have on the individuals concerned. Of course the personal involvement gave me a new perspective: all of us who went through those weeks acquired, I am sure, a deeper sympathy for those whose lives have been violated in this way.

The intercommunity violence and the cannibalism that took place at Gombe, however, were newly recorded and those events changed for ever my view of chimpanzee nature. For so many years I had believed that chimpanzees, while showing uncanny similarities to humans in many ways were, by and large, rather 'nicer' than us. Suddenly I found that under certain circumstances they could be just as brutal, that they also had a dark side to their nature. And it hurt. Of course, I had known that chimpanzees fight and wound one another from time to time. I had watched with horror when adult males, all inhibitions lost during the frenzy of a charging display, attacked females, youngsters — even tiny infants who got in their way. But those outbursts, shocking though they were to watch, had almost never resulted in serious injuries. The intercommunity attacks and the cannibalism were a different kind of violence altogether.

For several years I struggled to come to terms with this new knowledge. Often when I woke in the night, horrific pictures sprang unbidden to my mind — Satan, cupping his hand below Sniff's chin to drink the blood that welled from a great wound on his face; old Rodolf, usually so benign, standing upright to hurl a four-pound rock at Godi's prostrate body; Jomeo tearing a strip of skin from Dé's thigh; Figan, charging and hitting, again and again, the stricken, quivering body of Goliath, one of his childhood heroes. And, perhaps worst of all, Passion gorging on the flesh of Gilka's baby, her mouth smeared with blood like some grotesque vampire from the legends of childhood.

Gradually, however, I learned to accept the new picture. For although the basic aggressive patterns of the chimpanzees are remarkably similar to some of our own, their comprehension of the suffering they inflict on their victims is very different from ours. Chimpanzees, it is true, are able to empathize, to understand at least to some extent the wants and needs of their companions. But only humans, I believe, are capable of *deliberate* cruelty — acting with the intention of causing pain and suffering.

Meanwhile, oblivious of the concern they had caused me, the chimpanzees got on with their lives. And, for the Kasakela chimpanzees, retribution was at hand. After Sniff's death the victorious Kasakela males, along with their females and young, travelled, fed and nested without fear in their newly annexed territory. The size of their range increased from twelve to more than fifteen square kilometres. But this happy state of affairs did not last for long. The Kahama community, it seemed, had acted as a buffer between the Kasakela chimpanzees and the powerful Kalande community in the south. And now this community began to push further and further northward. A year after the Kasakela males had gained their final victory over Sniff, they were forced to begin a retreat. Again and again as they travelled in the area they had wrested, with such brutality, from the Kahama chimpanzees, Kasakela individuals encountered Kalande patrols. They began to move in the south with increasing caution and gradually their range began to shrink again.

Some dramatic encounters between Kasakela and Kalande groups were observed. Once, for example, Figan and four other males were routed by a larger group of Kalandeites and fled, in silence, back towards the north and safety. Two Kasakela males disappeared: first the strong young male Sherry and, the following year, old Humphrey. And, while we shall never know for sure, we thought it more than likely that they were victims of intercommunity aggression. After this, with only five adult males remaining, the Kasakela community not only continued to lose ground in the south, but also in the north where the large Mitumba community, seizing the opportunity, began to extend its territory southward. By the end of 1981, four years after Sniff's death, the Kasakela range was only about five and a half square miles — scarcely big enough to support the eighteen adult females and their families. I even feared that we might lose the whole community. Two of the more solitary and peripheral females who ranged most often in the south lost their infants, and, as with Sherry and Humphrey, we suspected that the Kalande males may have been responsible.

During the following year things came to a head. Four Kalande males actually arrived in camp and attacked Melissa. Fortunately — presumably because of their unfamiliar surroundings — it was a mild attack and her infant was unharmed. A few weeks later, when Eslom was out fishing one day, he heard Kalande males calling from the Mkenke-Kahama ridge just south of camp and, perhaps in reply, Mitumba males calling from the Linda-Kasakela ridge, just one valley north of camp. The Kasakela chimpanzees were being subjected to some of their own medicine. For several days after this they went about in silence. They even left one crop of succulent fruit hanging on trees by the Kakombe Stream — because, we thought, the rushing noise of the water would have made it impossible for chimpanzees feeding there to hear the approach of their 'enemies'.

Fortunately there were, at that time, an unusually large number of young males growing up in the Kasakela community. As time went on, they gradually began to spend more and more time away from their mothers, accompanying the adult males on their excursions to the north and south. These youngsters — Mustard and Atlas, Beethoven and Freud — lacked the strength and social experience to be of much use during an actual attack, but the sound of their calling and noisy displays, added to that of the four remaining senior males, may have given their neighbours the illusion that the Kasakela community was more powerful than it really was.

The danger was averted, and once again Kasakela patrols began moving south to Kahama, north beyond Rutanga. When they encountered males of the neighbouring communities, although both groups hurled challenges at one another as before, we saw no more dramatic chases. No more adult males, no other infants of peripheral females, disappeared. The status quo, it seemed, had returned.

Sons and Mothers

‹‹‹‹‹‹‹‹‹‹‹‹‹

PATROLLING THE BOUNDARIES is but one of the many duties that the young male chimpanzee must learn if he is to grow to be a useful member of society. His adult experiences will be very different from those of a female. Thus it is not surprising that the milestones along the path he follows towards social maturity are different from those that mark the route for the female. Some, of course, are shared — such as the weaning process and the birth of a new baby in the family. But the initial break with the mother and the first journeys with the adult males not only come much earlier for the young male than for the female, but are of far greater significance. For it is here that he must learn many of the skills that he will need as an adult. The young male must challenge the females of his community, one by one, and then, when all have been dominated, he must begin to work his way into the dominance hierarchy of the adult males. The way in which the young male tackles each of these tasks, and the age at which he passes from one milestone to the next, depend heavily on his early family environment and the nature of his social experiences. This becomes very clear if we compare the development of Fifi's sons, Freud and Frodo, with Passion's son, Prof.

As we have seen, even though Freud was a firstborn child, he enjoyed a relatively social infancy. Fifi's younger brother Flint was an important figure in Freud's first two years. Flint was fascinated by his small nephew, and Fifi was very tolerant, allowing him to play with and carry her precious infant when he was only two months old. Fifi's elder brothers, Faben and Figan, were often around too, and Freud developed friendly bonds with both of these high-ranking males. Thus, as Fifi herself had been, he was surrounded for much of his early life by supportive family members. He became, like his mother before him, self-confident and assertive in his interactions with his peers.

When Flint died at eight and a half years of age, unable to survive the loss of his old mother, Freud lost not only his principal playmate but also his adolescent male role model. Nevertheless he continued to enjoy a relatively rich social life. For one thing, even when old Flo was no longer around, a magnet to draw together the members of her family, Fifi still continued to spend time with her elder brothers. Freud always rushed to greet uncle Figan, jumping into his arms and even, on occasions, riding briefly on his back. This friendly relationship persisted after Figan had become alpha male. Moreover, Fifi was not only a sociable female, spending a good deal of time with other chimps, but after Flo's death — perhaps as a direct result of it — she became increasingly friendly with Winkle, a young female about the same age as herself. Winkle's son Wilkie was a year younger than Freud, and when the mothers were together their infants romped endlessly, using up some of their seemingly inexhaustible supply of energy. An only child demands a great deal of attention from the mother when she is the only other chimp around: thus the hours that Fifi and Winkle spent together, when they could feed or rest in peace, were beneficial to them as well as to their infants.

Of course, Freud went through the usual weaning depression — pressing close to Fifi when she rested, pestering her to groom him, desperately seeking reassurance in this new and unpleasant experience. And Fifi herself seemed upset during the early phase of weaning when, for the first time, the smooth, efficient coordination between herself and Freud, which had always been a characteristic of their relationship, was disrupted. Gradually both mother and son learned to cope with the situation, but Freud was still depressed when, for the first time since his birth, Fifi became sexually attractive again. Whenever his mother was mated by an adult male, Freud, in a frenzy of agitation, rushed towards the pair and, whimpering or even screaming, pushed at his mother's suitor. During the first and second of Fifi's pink swellings Freud seldom missed a single copulation; his distress and almost obsessive interference was reminiscent of Fifi's behaviour at the same age. Most youngsters appear to be less disturbed, although all interfere when their mothers are mated.

By the time Fifi's next infant was born, however, Freud had recovered from the stresses and strains both of weaning and of his mother's sexual popularity. He was enchanted by his baby brother Frodo, and as soon as Fifi would permit such liberties, Freud pulled Frodo from her arms, and sat grooming him or playing with him. He was almost always gentle, but there were many times when he made use of his young brother in order to get his way. If, for example, he was ready to move on before Fifi, and if, when he set off, she refused to follow, he would sometimes return, gather Frodo to his breast, and make off with his little brother. Sometimes this ploy worked and Fifi, with a sigh, got up and plodded after her two sons. But there were many occasions when she pursued Freud, retrieved her infant, and returned to get on with whatever she had been doing. There were times too when Frodo himself refused to play big brother's game and tottered back to his mother on his own.

There was a world of difference between the early experiences of Freud, the firstborn, and his younger brother. Even though Freud, in contrast to other first infants, had enjoyed a remarkably social environment, he had spent many long hours with only Fifi for company. And though she, like Flo before her, had been a playful mother, there had been countless occasions when she was too busy with her own concerns to pay attention to Freud. How utterly different it was for Frodo. He was never on his own with Fifi — his elder brother was always there. And Freud served, in turn, as playmate, protector and comforter, and role model.

It was different for Fifi, too, now that she had a second child. She was freed from the constant pestering of a bored infant, always wanting to be played with, wanting to be groomed. And she was freed not just *sometimes*, as when she had joined forces with Winkle after Flo's death, but *all* the time. She was able to sit, utterly relaxed, idly watching as Freud and Frodo played together. If she thought at all — and of course she did — she could think her own thoughts, uninterrupted. Even so, she remained playful herself and often seemed unable to resist joining in the games of her sons when she had nothing better to do.

Frodo was fascinated by almost everything that Freud did. He watched him carefully, then often tried to imitate what he had seen. When he was nine months old, for example, and still unsteady on his feet, he gazed wide-eyed as Freud did a noisy and impressive drumming display on the buttress root of a big tree, then did his best to do the same. But his coordination was not up to it — he lost his balance, tumbled down a slope, and screamed in terror — or was it frustrated anger? At any rate, his attempt at adult male behaviour ended in ignominious rescue by his mother. There was another time when Frodo, keeping very close to Fifi, watched Freud as he played aggressively with young baboons, chasing them, stamping his feet on the ground and flailing a large piece of dead wood. When all was quiet and the baboons had left, Frodo went over to the abandoned weapon, intent, no doubt, on demonstrating how fearsomely *he* could brandish it. But it was too heavy for him even to lift it from the ground.

Freud was very affectionate towards his young brother, and very protective too. When Frodo became adventurous and climbed beyond Fifi's reach, Freud often followed, seemingly to keep an eye on the infant. So that when, as often happened, Frodo 'got stuck' and whimpered in distress, Freud was close at hand to rescue him. When Frodo was about two years old he loved to play with baboons. Sometimes he got carried away and approached not only youngsters but even adults with his little displays. These adults occasionally became irritated with all his bristling and stamping and flailing of branches, and then they threatened him, slapping their hands on the ground and exposing their great canines. Frodo would scream in fear, and Freud was just as likely to rush to his rescue as Fifi was. Often, indeed, Freud stayed close by at such times, a self-appointed guardian.

While Frodo could hardly *rescue* his elder brother, he often showed concern if he was hurt or upset. When Freud was seven years old Fifi occasionally found it necessary to discipline him during feeding — if, for example, he tried to take a choice item that she had earmarked for herself. Twice when she mildly threatened her elder son he threw tantrums, hurling himself to the ground and screaming. Fifi ignored him, but little Frodo hurried over to his brother and embraced him, staying close by until Freud was quiet again. A year later Freud hurt his foot badly. He was unable to put it to the ground and for the first few days he travelled very slowly. Fifi typically waited for him when he stopped to rest, but sometimes she moved off before he was ready to limp after. Three times when this happened Frodo stopped, looked from Freud to his mother and back, and began to whimper. He continued to cry until Fifi stopped once more. Then Frodo sat close by his big brother, grooming him and gazing at the injured foot, until Freud felt able to continue. Then the family moved on together.

Most fascinating to watch was the interplay between Fifi and her two growing sons that led, for all three, to higher status in the community. Freud began the long struggle to intimidate the females of the community when he was seven years old. Charging towards and around them, he waved branches and hurled rocks — typical adolescent male behaviour. Initially he tackled the older juveniles and adolescents whose mothers were lower-ranking than Fifi. If one of these mothers turned on him — which was often the case — then Fifi would almost always back him up, threatening the female concerned, or even attacking her, for her ill-advised retaliation. Thus Freud's confidence grew and, as time went on he began to challenge the older females so that, more and more often, his 'victims' turned on their puny assailant and chased him off, or even beat him up. Fifi, because she almost always went to his defence, was increasingly drawn into conflicts with the other females.

There were times when Freud aimed too high. Once, for example, he had the audacity to threaten high-ranking Melissa, and she thrashed him soundly for his rashness. Fifi, although younger and lower ranking than Melissa, had, like Flo before her, a staunch and fearless nature. In response to Freud's anguished screams she rushed up, her hair bristling, uttering fierce waa-barks of threat. Melissa at once turned from Freud to Fifi, and the two mothers fought, grappling and rolling over and over. Freud ran behind them, uttering high-pitched and futile waa-barks of his own. Unfortunately for Fifi, Melissa's adolescent son Goblin was nearby and, hearing the screams of his mother, he charged up, attacking Fifi and chasing her — and Freud — away.

But all the time Freud was growing bigger and stronger and, as levels of the male hormone, testosterone, increased during puberty, he became more aggressive too. By the time he was nine years old he was

Fifi and her first-born, Freud. Chimpanzee infants seem to have unlimited energy. Fifi is normally responsive when Freud tries to initiate play—but not always.

As Frodo, Freud's younger brother, got older, he became an even better playmate.

Fifi, like Flo, often plays vigorously with her children. Here she enjoys a romp with Frodo.

Prof throws a weaning tantrum, utterly ignored by his mother, who is licking the salt block.

Ajax mounts Moeza. She has lost her mother and ignores the whole performance.

Quisqualis threatening Pom. Note Prof's hand holding onto his mother for reassurance as she licks salt.

Wunda and a young female of Camp Troop.

A four-week-old baboon infant with its mother.

Baboons grooming.

Melissa stares down at her one-day-old son, Goblin.
(Hugo van Lawick)

Goblin was fascinated by pink bottoms from an early age. He and adolescent Fifi are practising.
(P. McGinnis)

From the time when he began to leave his mother and travel with the adult males, Goblin hero-worshipped Figan. And Figan was very tolerant of his small follower. (C. Packer)

Goblin, in foreground, as top-ranking male. Behind sits the much larger, but clearly subordinate, Satan. They are replying to pant-hoots from over the valley. (Kenneth Love, © NGS)

Opposite, top: Patti chasing Tapit around Goblin—who was almost certainly his father.

Opposite, bottom: Goblin with his sister Gremlin, who cradles Getty.

Jomeo, as an adolescent, showed strange outbursts of screaming when he came to camp alone and was given bananas. Indeed, he sometimes screamed so loudly that his throat cramped and he could not swallow. (Hugo van Lawick)

Jomeo, his face swollen owing to an abscess of his tooth.

Below: Jomeo and his brother Sherry. For years they were all but inseparable.

able to support his mother in *her* altercations. When Fifi once became involved in a fight with the high-ranking Passion, both Freud and Pom joined the skirmish, in support of their respective mothers. But Freud was able to chase Pom away, and then he returned and hurled a rock at Passion. This startled her and allowed Fifi to win the fight. And so, as the years went by, both mother and son gradually raised their social standing.

Meanwhile young Frodo was growing up too. Secure in the knowledge that, if things went wrong, Fifi or Freud — or both of them — would surely come to help him, he began to challenge community females at a very early age. After all, he had been watching Freud, learning from him — and, indeed, 'helping' him — for years. Again and again, as Freud had threatened some wretched female with his swaggering displays, little Frodo had joined in: his every hair bristling he had bounced and stamped about on unsteady legs, swaying tiny branches, looking for all the world like an animated figure from a Disney cartoon.

Frodo was just five years old when he began to challenge some of the females on his own. Of course, he was still very small, but he quickly learned that the judicious use of rocks as weapons tremendously enhanced the effectiveness of his threats. Very soon he developed a reputation as a prize thrower. Many young chimps throw rocks during intimidation displays — it was a characteristic component of Freud's performances and it is more than likely that Frodo, initially, was imitating his elder brother. But Frodo perfected the throwing technique, and, in a very short space of time, many of the younger and lower-ranking females came to fear this precocious young male and hastened away when he swaggered towards them, rock in hand. Frodo scored direct hits rather more often than other stone-throwers, not so much because his aim was better, but because he approached to within a couple of feet before hurling his missiles. He developed other unpleasant techniques as well.

I remember vividly an incident that happened when I was following Fifi, Little Bee, and their families. Suddenly Little Bee, gazing up the steep slope, began to utter small screams. And there, some yards above us, I saw Frodo just starting a swaggering display, hair bristling, rock in hand. He hurled it towards us but it fell harmlessly between Little Bee and myself. It was not clear whether Little Bee or I had been the intended victim — Frodo had always considered that I was merely another female, to be dominated along with the rest. Next he began pushing at a huge rock. It was much too big for him to lift, but he could — and did — set it rolling down the slope. It gained momentum rapidly as it bounded towards us, ricocheting erratically from one tree trunk to another. Had it hit any of us we could well have been knocked out, if not killed. And then, even as I was wondering which way to run, Frodo set a second rock in motion. By the time he had started to push at a third we were all running for our lives — not only Little Bee and myself, but Fifi too. Fortunately Frodo did not make a habit of this type of bombardment, although he continued to throw stones and small rocks for years.

One of the most important milestones in the life of a young male is when he begins to travel away from his mother with other members of the community. The severing of apron strings is far more necessary for a young male than for a young female. She can learn most of what she needs to know for a successful adult life whilst remaining safely in her family setting. Not only can she watch her mother and her mother's friends caring for their infants, but she can actually handle them herself, gaining much of the experience which she will need later when she has a baby of her own. And she can learn, during her mother's 'pink days', a good deal about sex and the demands that will subsequently be made of her in that sphere.

The young male has different things to learn. There are some aspects of community life that are primarily, though not entirely, male responsibilities — such as patrolling, repelling intruders, searching out distant food sources, and some kinds of hunting. He cannot gain adequate experience in such matters if he remains with his mother. He must leave her and spend time with the males. Freud had been fascinated by the big males throughout his infancy. From the time he could walk he had been quick to totter up to greet any males who joined his mother, and often, too, had followed a short way when they left. I remember Freud stumbling after Humphrey once, as he set off after a grooming session with Fifi. His mother, not at all wishing to leave, followed and tried to retrieve him, but he protested vigorously, whimpering and clinging tightly onto the vegetation. After a few attempts, each one provoking increased resentment, Fifi gave in and trailed along behind as her son continued to follow Humphrey. Presently, though, he got tired, climbed onto her back, and made no protest when she moved off in the direction of her choice.

Freud had always been eager to join in the fun whenever he heard the calling of chimpanzees gathered in excited, noisy groups. I remember one occasion, when he was just four years old. We had had a peaceful morning, just the three of us on our own. At midday Fifi rested, stretched out on the ground, while Freud, ever active, played in the branches above. Suddenly, on the far side of the valley, there was an outburst of excited pant-hoots and screaming. Certainly some of the adult males were there — the voices of Figan and Satan, Humphrey and Jomeo were easy to recognize — and we could hear females and youngsters as well. Freud listened intently, then joined in with his high-pitched infant pant-hoots and Fifi sat up and called as well. Swinging down, Freud at once set off in the direction of the big group. But Fifi did not move and, after travelling some ten yards, Freud looked back, then stopped and whimpered softly. But Fifi ignored her son's plea and lay down to continue her rest. Disappointed, he moved back and sat beside her, raising one arm in a request for grooming.

Five minutes later the group called again. As before Freud joined in eagerly, this time running along the ground and stamping his feet in a small display. Again he set off along the trail towards the excited calling, longing to be part of it, to join his peers in their games. But still Fifi made no move to follow. This time Freud went a little further before he paused and looked back. Nor did he return, but remained standing some fifteen yards along the track — just before it made a sharp bend that would take him out of Fifi's sight. Gradually his soft whimpers increased in frequency and volume until he was crying loudly.

And then, either because of Freud's entreaty, or just because she felt like joining the fun herself, Fifi got up and followed her son along the trail. Ten minutes later they were part of the noisy, exuberant group. Fifi, with soft grunts of pleasure, climbed to feed on the juicy figs that had attracted more than half of the community members to the feast. Freud, beside himself with excitement, raced to join a wild play session with other youngsters.

One very clear indication of increasing independence in the young male is the frequency with which he joins gatherings of this sort without his mother. Sometimes the chimpanzees get together in these large and noisy groups in order to feast on some abundant and delicious crop; sometimes the magnet is a sexually popular female. The gatherings usually last for a week or more, with chimpanzees arriving and leaving at different times. In many ways they are the hub of chimpanzee social life, giving community members the opportunity to meet and interact with each other — to play, groom, display, make a noise. Often, particularly when several pink females are present at the same time, there is almost a carnival atmosphere.

Fifi, with her social disposition, joined many gatherings throughout Freud's infancy and childhood, so that he gained much social experience and learned (often the hard way) to make himself scarce when the big males were tense and the threshold for aggression low. As the years went by, Freud's self-confidence in such situations increased: by the time he was nine years old he was joining gatherings without his mother quite regularly. And Frodo did so at an even earlier age — provided that his big brother was nearby to provide reassurance in times of stress. Indeed, when he was only five years old, Frodo actually spent several nights in succession away from his mother, travelling with the adult males — and Freud.

Prof's childhood was very different from that of Freud and even more so from that of Frodo. Although Passion was considerably more attentive and less harsh with this her second child, she could not begin to compare with Fifi in terms of affection and solicitude, tolerance and playfulness. Moreover, with the passing of the years she had become increasingly asocial — the big groups of chimps that had gathered in camp for bananas during Pom's infancy were a thing of the past. And Passion had no friend, such as Winkle, with whose infant Prof could play. He did, of course, have an older sister but even though, after getting over her weaning depression, she began to show more interest in her young brother, she never played the role in his life that Freud had played in Frodo's or Flint, before he died, in Freud's.

Prof, therefore, had less opportunity for social interactions of any kind than Freud and Frodo did. Perhaps because he played with other youngsters less often than they, he lacked confidence when he did play. He hardly ever stood up for himself when a game got rough even though, if he did get into trouble, Pom as well as Passion usually helped him out. But probably the most significant difference in the early social experiences of these three young males was the fact that Prof had far fewer opportunities to interact with adult males.

For Prof, as for his sister before him, weaning was a time of despair, but because he was a male he was far more aggressive in his misery than Pom had been. He threw violent tantrums, screaming and tearing at his hair, hurling himself to the ground. In most families, tantrums elicit an immediate response from the mother. Frodo, spoilt child that he became, had also thrown violent tantrums. In his case, I think, they were due more to rage at not getting his own way. Always Fifi had reached out to him, trying to draw him close. If, as was so often the case, he had then hurled himself to the ground, pulling away from her conciliatory gesture, she had usually taken him into her arms and held him there. And, however violent had been his rage, Frodo had always calmed after a while, perhaps intuitively understanding his mother's message: 'You can't have milk (or ride on my back) but I still love you, anyway.'

But hard-hearted Passion often ignored Prof's tantrums altogether. This, of course, was yet another form of rejection, and Prof became increasingly distressed as a result. Screaming loudly he would rush off through the undergrowth or hurl himself down some slope. Once he actually tumbled backwards right into a stream — and young chimpanzees are frightened of fast-moving water. Even then, when his screams of frustration must surely have turned to screams of fear, Passion ignored her son. This troubled period in his young life did little to boost Prof's already minimal self-confidence! However, unlike Pom, Prof recovered from his weaning despair before the birth of his infant brother, Pax, and, like Freud, he was fascinated by his new sibling, more so than Pom had been by hers.

Prof was about the same age as Freud when he was first seen to challenge a female. But whereas Freud, having once embarked on the task of dominating the females, repeated his displays more and more

often, Prof's performances were few and far between. And they lacked
the determination and vigour that characterized Freud's efforts and,
later, Frodo's. Indeed, Prof's second attempt ended somewhat ignominiously when his 'victim' reached out, grabbed his neck, and tickled
him until his bristling aggression ended in laughter.

Prof, as an infant, clearly longed to spend time with the big males
just as Freud and Frodo had. But if he set off after one of them Passion
never followed and so quite soon he gave up trying to persuade her.
Moreover, because Passion avoided the big groups that Fifi and other
sociable females found so exhilarating, Prof often seemed ill at ease
on those occasions when he did find himself in such a gathering. And
so, lacking Freud's and Frodo's self-confidence, Prof was still spending
almost all his time with his mother when she died — at which time
he was almost eleven years old.

There can be little doubt but that differences in behaviour observed
in Freud, Frodo and Prof stem, in large part, from the different personalities and child-raising techniques of their mothers. Of course,
there are genetic differences also between these three young males:
some temperamental differences surely derive from heredity rather
than experience. Sometimes, though, one can trace the onset of an
unusual behaviour to a particular traumatic incident that occurred in
early childhood. When Prof was two years old, for example, he was
attacked by an adult male colobus monkey during a hunt. Passion was
just sitting and watching, holding Prof, when suddenly one of the
colobus males, enraged, leapt at and attacked her. She was quite unharmed: Prof had one toe bitten right off.

That experience, both painful and frightening, apparently left Prof
with a deep-rooted fear of monkeys. Most young males begin to hunt
when they are mere juveniles. Freud caught his first monkey (which
Fifi took from him) when he was only six years old. Prof was not
observed to hunt monkeys at all until he was eleven, and even then it
was in a half-hearted manner. He has never been observed, by us, to
catch one. Interestingly, Prof was also terrified of baboons as a child.
He showed none of the swaggering, bristling, aggressive play with
young baboons that we saw so often in Freud and Frodo. If a large
male baboon approached him, during feeding for example, he whim-

pered in fear and hid behind Passion. Thus it seems that his fear of colobus monkeys may have generalized into fear of all monkeys and baboons. Of course, there is always the possibility that there was some equally traumatic interaction with baboons that led to this second childhood fear. Certainly there would have been many opportunities for such an event to take place.

Baboons

‹‹‹‹‹‹‹‹‹‹‹‹‹

T HE INTERACTIONS between chimpanzees and baboons, as observed at Gombe, are more varied and more complex than those between any other two species in the animal kingdom — with the exception of our own interactions with other animals. Chimpanzees and baboons sometimes compete aggressively for food. Young baboons may be captured, killed and eaten by chimpanzees. The young of the two species sometimes play together — and young chimpanzees may even groom and try to play with adult baboons. Finally, they understand many of each other's communication signals, and sometimes this results in what amounts to a joint effort to intimidate and repel a predator.

There are more baboons than chimpanzees at Gombe for, while the number of individuals in each social group — the baboon troop or the chimpanzee community — is about the same, averaging fifty over the years, there are some twelve troops of baboons crowded within the range of one chimpanzee community. This means that it is rare indeed for a day to go by without an encounter of some sort between individuals of the two species. For the most part these meetings are peaceful: often the chimps and baboons simply carry on with their own pursuits and seem to ignore each other altogether. They do, of course, utilize many of the same food resources. The food supply at Gombe is, for most of the year, more than adequate for the requirements of both chimpanzees and baboons, in which case there is no need for them to squabble. On occasion, individuals of the two species feed peacably in the same tree. At other times there may be varying amounts and intensities of aggression. It is during the dry season, from June to October, when food is sometimes in relatively short supply, that one sees the most aggressive competition between the two primate species. When a baboon troop arrives near a tree where three or four chimps are feeding, and its members, one after the other, climb into the branches, the chimps tend to become increasingly nervous. Moving rapidly from place to place, they stuff food into their mouths more and more quickly, then they usually leave. But not always — sometimes, even when they are heavily outnumbered, the chimps do not give up so easily. It depends on the age, sex and personalities of the individuals present. Some chimps are far bolder than others in situations of this sort — and there is no question but that the baboons recognize them. I well remember an occasion when Goblin, Satan and Humphrey were peacefully feeding on figs, and D troop baboons arrived and climbed up, more and more of them, to share the feast. Led by Goblin the three male chimps charged the baboons again and again. There were violent skirmishes in the branches, chimps and baboons screamed and roared, the quiet of the morning was shattered. It was only after twenty minutes that the chimpanzees finally decided to call it a day. Even then they made an impressive exit, uttering loud roaring hoots and charging through the baboons who were feeding on the ground, scattering them, screaming, in all directions.

Some chimpanzees are far more fearful than others in their interactions with the baboons — and the baboons, knowing this, react accordingly, taking liberties with some chimpanzees they would not take with others. Likewise, the chimpanzees recognize that certain adult male baboons are not to be trifled with. Walnut, for several years alpha male of Camp troop, invariably struck fear into the hearts of the staunchest chimpanzees. And rightly so, for he sometimes appeared to go berserk, charging hither and thither through a peaceful group of chimpanzees, uttering the fierce roar-grunts that sound every bit as frightening as the coughing roar of a leopard, until one and all had fled.

Nevertheless, despite the occasional dramatic confrontation over a valued food resource, most disputes are settled peacefully, with nothing more vigorous than a mild threat gesture from one side or the other. Competition is minimized by the fact that baboons have a more catholic diet than do chimpanzees. They eat a greater variety of stems and seeds and blossoms. They spend hours digging for roots and little nodes in the dry season, when food is scarce. They turn over rocks in the streams and on the mountain slopes looking for crabs and insects. Their incredibly strong jaws enable them to crack open the small rock-hard kernels of the oil-nut palm fruits. The Gombe chimpanzees, die-hard conservatives that they are, seldom show interest in any food item that is not part of their traditional diet. Except for the infants — sometimes they seem fascinated when they see baboons feeding on something different.

I vividly recall one incident. Pom was resting as her two-year-old son, Pan, played nearby. A number of baboons were peacefully foraging in the vicinity, and one of them, the adult male Claudius, sat near the two chimps. Pan moved closer and watched with a wide-eyed stare as Claudius picked up a palm-nut kernel, placed it between his molars and, pressing up on his lower jaw with one hand, cracked the shell. He extracted the nut and let the two halves of the now hollow kernel drop to the ground. Pan, keeping his eyes fixed on the baboon's face as though trying to gauge his mood, very cautiously approached, reached out and seized a piece of shell. Overcome with his daring, he hurried back to Pom and, holding her hair with one hand, carefully examined and licked his prize. Claudius, by this time, had selected another fallen kernel and Pan watched, with similar fascination, as that was cracked open too. Then, this time with greater confidence, Pan again approached the baboon and picked up the discarded shell.

If the food had been something that Pan could easily have obtained for himself, like a berry growing on a bush, I am sure that he would have picked and eaten one. In that way a new feeding tradition could have been started, learned originally from the baboons. But the rock-hard palm nut posed too difficult a problem for an infant chimpanzee.

The rich, nutritious outer flesh of the fruit of the oil-nut palm is, however, a staple for chimpanzees and baboons alike as the trees ripen one after another throughout the year. Each palm offers only one or two feeding places and, when food is scarce, there may be fierce competition for access to the clusters of red fruits. I remember one time when, as I followed Fifi through the forest, she suddenly paused and, hair bristling, stared up into a tall palm. A moment later she rushed up the trunk and, as she neared the crown, a very small juvenile baboon, screaming in fear, leapt away along one of the fronds. I watched, holding my breath, for I thought that Fifi was trying to catch the youngster — even though in twenty-five years we had never known a female to take part in a baboon hunt.

But Fifi wanted only to gain access to the one cluster of ripe palm fruits up there. As she settled down to feed, uttering soft grunts of delight, her hair gradually sleeked. Meanwhile, however, the small baboon was in a predicament. Perhaps he too had mistaken Fifi's aggressive mien for a predatory interest in himself. At any rate, he seemed determined not to venture anywhere near the female who had given him such a scare. Clinging to the very tip of the frond he looked around vainly for a way to escape. His weight was not sufficient to pull the frond all the way down so he hung some ten feet out from the trunk of the palm. There were no convenient branches nearby into which he could leap. For over three minutes he remained thus suspended. And then, gradually gaining confidence, he very quietly and cautiously climbed back up the frond towards Fifi until he could reach a neighbouring frond. He worked his way, oh so silently, around the palm, from frond to frond until, at last, he could leap into a nearby tree and make his getaway.

Tall palm trees, with crowns emerging from the surrounding canopy, have occasionally served to trap baboons on the relatively rare occasions when they are hunted by chimpanzees. If a hunter manages to creep stealthily up the trunk, while others wait on the ground below, the intended prey may find it difficult to escape. Once, for example, six male chimpanzees, travelling in the south of their range, came upon a female baboon with a very small infant feeding, quite by herself, in a palm tree. She was not a member of any of our study groups and we did not know her by name. Figan, who was in the lead, grinned when he saw her, squeaked softly, and reached to touch Satan. All six males stood gazing up, their hair bristling. When the baboon noticed them she stopped feeding and, almost at once, began to show signs of distress, giving soft fear calls and backing away to the other side of the palm. Jomeo, moving slowly, climbed a tree close to her palm until he was level with the baboon and about five yards away. As he stopped and stared at her she began to scream loudly, but apparently no other baboons were within earshot. Certainly none appeared, then or later.

After a tense two minutes, Figan and Sherry climbed deliberately into two other trees. One hunter was now stationed in each of the trees to which their victim could leap. The other three chimpanzees waited on the ground. Suddenly Jomeo leaped over into the baboon's palm. The baboon made a huge jump into Figan's tree. It was easy for him to seize her and pull away the tiny baby. He killed it with a quick bite into its head. And then, as the mother watched and called out hopelessly from a neighbouring tree, the six hunters shared the carcass.

Because we also study the baboons at Gombe and know the members of five troops by name, along with their fascinating life histories, it is always traumatic when they are killed and eaten by chimpanzees. Yet there is an undeniable sense of excitement when such a hunt begins and a mounting feeling of suspense among us. More often than not baboon hunts fail. Had that female's troop been nearby when Figan and his friends arrived on the scene, things would have gone very differently. Baboon males are fierce when roused, and as soon as they hear the distressed screaming of an infant or its mother they race to the rescue, roaring, lunging and hitting at any chimpanzees in the vicinity. Adult females join in too, at least adding to the commotion with their screeches of fear and rage. In the face of such mobbing many attempted hunts are abandoned and the chimpanzees flee. Indeed, it always amazes me that, given the fury of the defence, chimpanzee hunters ever manage to seize and kill a victim. Even more amazing is the fact that on all occasions when we have observed successful hunts, the chimpanzees, though they may be seized and held to the ground by infuriated male baboons, have never been actually injured by them. Yet baboons will attack a leopard who hunts their young, and may wound it so severely that it later dies. It seems that the chimps, perhaps by virtue of their ability to hurl sticks and rocks at their opponents, have established themselves as the dominant species. They have, in effect, bluffed the baboons into believing them to be stronger and more dangerous than they actually are.

Baboons are hunters too — there are records of meat eating from almost all parts of their range across Africa. At Gombe they most often catch the fawns of bushbuck during the birth season, when the mothers leave their young pressed to the ground in areas of tall grass. Because baboons spend more time than chimps searching for food in such places, and because they spread out when foraging, they are more likely than the chimps to come upon the hidden fawns.

Once a baboon has captured prey there is usually a good deal of aggression as the captor, trying to feed, is harassed by his companions. Often, during these skirmishes, the carcass is taken over by a succession of adult males. All this makes for a lot of noise, a cacophony of screams and barks and roars. If chimpanzees hear a commotion of this sort they usually stop whatever they are doing and race towards the sounds. Then follow amazing acts of piracy.

I have already described the encounter between Gilka and the male baboon Sohrab. She, small and weak, failed to take over the prey. Other females have been more successful. One of the most dramatic incidents was described by Hilali. He was following Melissa and her two offspring: her five-year-old son Gimble and her ten-year-old daughter Gremlin. A sudden medley of sounds from the baboons of D troop, who were foraging nearby, brought the chimpanzees, who had been quietly grooming each other, instantly to their feet. With grins of excitement they embraced briefly, then raced together towards the uproar. A few moments later they came upon the adult baboon Claudius tearing at the meat of a freshly killed fawn. Three other males were threatening him, slapping their hands on the ground, showing their canines and the whites of their eyelids as they yawned, uttering fierce-sounding roar-like grunts.

Melissa and Gremlin slowly moved closer, watching as Claudius dragged his prey along the ground. Then, as he paused to tear off another mouthful, they charged towards him uttering loud barks of threat and waving their arms. When the baboon retaliated, roar-grunting and lunging fiercely in their direction, Melissa stopped. She gave a few small whimpering sounds, then seized a thick dead branch and, hair bristling, hurled it towards Claudius, who leapt aside. Quickly following up her advantage, Melissa charged again, this time swaying the vegetation wildly, leaping up and down, gradually moving closer.

Suddenly Claudius dropped his prey and lunged at Melissa, hitting her and, Hilali thought, biting her arm. Melissa fought back, barking loudly, flailing her arms and hitting out at her powerful adversary. At this point the other male baboons, seizing their opportunity, converged on the prey, and Claudius was forced to turn from Melissa to retrieve his meat. Melissa watched for a few moments and then began another wild display. Gremlin joined her mother again and once more they charged Claudius as a team. He held his ground but began to feed in a frenzy, tearing pieces of meat from the rump of the fawn. Melissa watched and, from time to time, shook vegetation and whimpered.

After five minutes she began to display again, even more wildly this time. Claudius seized the carcass in his mouth and tried to drag it further away, but it got tangled in the undergrowth. After tugging desperately and in vain, he tore off a large piece and ran away with it. But when Melissa rushed to the prey and seized a front leg he returned and grabbed the other end. Amazingly, despite his terrifying roar-grunts and the close proximity of those gleaming canines, Melissa, screaming loudly, hung on. And Gremlin, who had rushed up a tree when Claudius grabbed the prey, soon swung over above the scene of conflict and began to wave and shake branches just above her mother, adding to the confusion. And then Melissa, still hanging onto the carcass for dear life, started to climb up towards her daughter. Suddenly the baboon seemed to lose his grip and Melissa, quickly flinging the carcass over her shoulder, climbed higher. Then, even as Claudius, roar-grunting, leaped after her mother, Gremlin seized a dead branch, broke it off, flailed it wildly at the baboon and then hurled it at him. He managed to dodge this missile and again lunged towards Melissa. But at this point she seemed suddenly to lose her fear of him and, eleven minutes after the conflict began, started to feed quite calmly on the stolen meat. She shared with Gremlin and with young Gimble who had watched the entire incident from safe vantage points in the trees. For a while Claudius sat close by and continued to threaten, but when two other female chimpanzees arrived to share the meat he gave up and climbed down to join the other baboons who were milling about below the tree, searching for fallen scraps.

How is it that a female chimpanzee, with her relatively short, blunt teeth can face up to a fully adult male baboon with canines twice as large and powerful as hers — and win? Is it her swaggering display that accomplishes this seeming miracle? The bristling hair, the wildly shaken branches, the upright posture that is so often assumed? Or is it the use of weapons — the branches that may be flailed or thrown? Probably a combination of these things, together with the fact that if other male baboons are present they will not help the possessor of the meat but rather try to steal his prey, distracting his attention from his chimpanzee adversary. Male baboons, though they cooperate in the defence of their troop from rival males, have not been observed to cooperate during hunting, nor do they share the prey when a kill is made.

Only once have we observed a baboon stealing meat from a chimpanzee. This was when Passion had killed a wounded hawk — a large bird with a wingspan of at least three feet. As she sat feeding, sharing with Pom and Prof, Hector, a Camp troop baboon, approached. He sat nearby, watching. Presently young Prof, seven years old at the time, managed to persuade his mother to part with a whole wing. Uttering loud grunts of delight he moved a few yards away to feed. Seizing his chance, Hector raced towards Prof, seized the wing, and rushed off with it, leaving Prof to throw a violent tantrum, almost choking in his rage.

The sounds made by baboons who have captured prey are very similar to the uproar heard during some other aggressive incidents: occasionally chimps make a mistake and race up to a baboon troop, apparently expecting a mouth-watering feast, only to find that fierce competition has broken out over, for example, a female in oestrus. Not very interesting to a chimp — although an adult male will often watch with the expression of a connoisseur, as a fully swollen female baboon walks past. If she pauses and turns her rump to him, in the typical primate submissive 'present' posture, he may reach out and touch, or at least sniff, her bottom — as he would if she were a chimpanzee. Infant and juvenile males show even more interest in the pink swellings of female baboons and may actually attempt to mate with them. Once this led to the most incredible communication sequence I have ever seen between non-human animals of different species.

The actors in the drama were seven-year-old Flint and the adolescent female baboon Apple of Beach troop. Flint, clearly, was sexually stimulated by the sight of Apple's small rosy swelling. To attract her attention he used postures and gestures typical of male chimpanzee courtship: he sat and looked towards Apple, his thighs splayed, his penis erect, and he shook a little branch with quick jerking movements of one arm. With the exception of the erect penis, a male baboon does none of these things — he simply approaches the female of his choice and gets on with the job. Apple, however, seemed to understand quite well what Flint wanted — probably she wanted it too. She approached and presented herself for copulation. She did this in the manner of her kind — she stood squarely in front of Flint, facing away from him, looked back over her shoulder, and held her tail to one side. But this is not how a female chimpanzee offers herself to her male — *she* crouches close to the ground. Flint looked at Apple, perplexed. He shook his branch again. And then, seeing that this was not effective, he stood upright, placed the knuckles of his right hand on her rump, at the base of her tail, and pushed down. To my amazement, Apple flexed her legs — but only a little. Flint stared at Apple, shook his branch again, then repeated the pushing exercise. Apple flexed her legs just a little more. Now it seemed that Flint was prepared to meet her half-way. The male chimpanzee normally copulates in a squatting position, his body more or less upright, one hand often resting lightly on the female's back. By contrast, the male baboon grips the female's ankles with his feet, grasps her around the waist with both hands, and, thus elevated, gets down to business. Flint gripped Apple's right ankle with his right foot, held onto a sapling with the other foot, and actually achieved intromission.

This sequence in its entirety was incredibly sophisticated: Flint and Apple each seemed to understand just exactly what the other wanted, and adjusted their behaviour accordingly, even when this meant doing things not normal for either of them.

Sometimes young male baboons get turned on by adolescent female chimps, grasp their ankles, and try to attain intromission. But we have never recorded a sequence that was as sophisticated as that observed between Flint and Apple. The most amusing incident occurred when

Miff's daughter Moeza was nine years old. She was barely swollen, and anyway she was in no mood for sex play, as she had temporarily lost her mother and, from time to time, was crying softly to herself. When young Hector of Camp troop approached and mounted her, three times in succession, she simply stood, looking depressed and forlorn, and completely ignored his unsuccessful efforts to mate her.

Chimpanzees, clearly, understand and may respond appropriately to many of the postures, gestures and calls of the baboon communication system — signals given in friendliness and threat, submission and sex. Equally, the baboons understand similar messages conveyed by the chimpanzees. Individuals of each species are alerted by the alarm calls of the other — indeed, they pay attention also to the alarm calls of various kinds of monkeys and even birds. This is commonplace in nature — the news of some danger, such as a prowling leopard, is broadcast by the individual who discovers it and members of other species have learned to recognize the call for what it is. This is highly beneficial to the potential victims of a carnivore and, presumably, most frustrating for the hunter.

One day, as I followed Fifi and her family through the forest, we heard the loud, insistent alarm calls of the Camp troop baboons on the other side of the valley: 'waa-hoo! waa-hoo! waa-hoo!' First one baboon shouted out his news, then the message was repeated by more and more of his companions. Shrill juvenile voices and the deeper calls of females joined the gruff chorus of the big males. Fifi paused, with Flossi perched on her back, Fanni a few steps behind, and stared towards the commotion. After a few moments Fifi decided to investigate. Turning from the trail she had been following she plunged into the tangled undergrowth of the lower slopes. Struggling to keep up I crawled and wriggled behind. Soon we crossed the stream and started up the opposite slope. As we got closer Fifi kept stopping to peer ahead through the vegetation. Suddenly there was a rustle close by. Fifi turned and, with a wide grin — of fear and excitement or both — reached her hand towards the dark shape of another chimpanzee, half-glimpsed in the undergrowth. It was Goblin, hair bristling — and he too grinned as he reached to touch her hand. Reassured by the contact the two moved on. Presently I was aware of other silent forms moving

alongside, all heading towards the place where the baboons confronted the unknown danger.

The first baboons came into sight, perched on low branches and
staring down at the forest floor. Every so often one started a new series
of 'waa-hoo! waa-hoo! waa-hoo!' The chimps — and there were
about eight by now — climbed into the trees and also peered down
through the leaves. What was there? I felt decidedly uncomfortable
until I located a tree up which I too could climb should the need arise.

Suddenly Fanni gave a soft 'huu' — a sound signifying wonder,
puzzlement, just a little fear. Fifi moved closer and started down, following the direction of Fanni's gaze. Then she too gave a 'huu' followed, almost at once, by the spine-chilling 'wraaa', the chimpanzee
alarm call. This acted as a signal to the other chimps and I found
myself in the centre of a fearsome chorus. The males, hair bristling,
suddenly began spectacular arboreal displays, leaping from branch to
branch, shaking and swaying the vegetation.

As yet I had seen nothing, but all at once, as Satan leapt, with a
fierce call, almost to the ground, I saw it too, or part of it: an extremely
large python, as big around as a man's thigh. So perfect its camouflage
that I would never have seen it had not Satan's display set it moving,
slowly, through a patch of sunlight.

For the next twenty minutes the chimps and baboons hung around.
No longer scared, they were curious, fascinated. First one, then another, would move down, closer and closer, only to leap back, with
startled exclamations, if the python moved. But gradually, as the snake
glided into dense undergrowth and disappeared from sight, the spectators lost interest. The baboons left first, and then, in twos and threes,
the chimps wandered off as well.

We have no evidence that pythons at Gombe have ever killed young
chimps or baboons, but theoretically this would be possible. There are
tales of pythons catching, suffocating and swallowing very large animals. By alerting each other to potential danger of this sort the chimps
and baboons may well be of service to each other from time to time.

Of all the interactions between chimpanzees and baboons, it is,
perhaps, the exuberant play sessions of the youngsters that are the
most fascinating to observe. Sometimes an unusually close relation-

ship — a friendship really — develops between a young baboon and a chimp and they will seize every opportunity to play together. The first such friendship I observed was in the early sixties between Gilka and the young female baboon, Goblina. Whenever Gilka's mother was near Goblina's troop the two youngsters would seek each other out and start to play, gently tickling each other with their fingers or with nibbling, nuzzling movements of the jaws. Their play was accompanied by soft laughing. Sometimes one of them would briefly groom the other. Sadly, Goblina's first baby was killed and eaten by chimpanzee hunters. Gilka had no part in that incident, but I suspect she would have begged eagerly for a share of the meat had she been around at the time. There are many small-holding farmers who have sat together to eat the pig who was, for a while at least, almost part of the family. Gilka would have had less reason to reject the flesh of Goblina's infant.

More recently a similar, though much less gentle, relationship developed between juvenile Freud and the young baboon Hector. Again and again the two would rush towards each other and wildly tumble together, Freud, the smaller, sometimes laughing almost hysterically as the session became increasingly rough. I never saw Gilka and Goblina being aggressive to each other, but the play between Freud and Hector frequently degenerated into aggressive chasing, even fighting. Hector usually came off best, and Freud, screaming, would run to Fifi for comfort. But when they met next time Freud was as eager to play as ever.

For the most part chimp–baboon play comprises chasing and brief episodes of sparring. The chimps, particularly the young males, tend to display many aggressive patterns during these sessions, stamping their feet on the ground, flailing branches, and throwing rocks. Often these play sessions end with the baboons running off screaming. Sometimes the routed baboons approach one of the adult males and then, feeling safe, turn and threaten their too rough playmates. Occasionally adults of both species are drawn into these childhood squabbles and start hurling abuse at each other: the chimps wave their arms, flail branches and utter waa-barks; the baboons roar-grunt, flash their eyelids and display their fearsome canines while lunging towards their opponents. All this, however, is usually but 'sound and fury, signifying nothing' and, after a lull, the play often starts up again.

Perhaps the most extraordinary incident I have ever observed between a chimpanzee and a baboon was one involving Pom and Quisqualis of Camp troop. Pom, from childhood onwards, had typically shown a marked lack of respect for adult male baboons and their powerful canines. On this particular occasion, when she was about ten years old, her behaviour seemed to verge on sheer lunacy! The incident took place in camp, during the days when I occasionally set out a mineral lick that was enjoyed by chimpanzees and baboons alike. Passion and her family had been licking for some time when the baboon Quisqualis arrived and tried, most vigorously, to displace the chimpanzees. Many chimps would abandon their licking in the face of just one serious threat from an adult male baboon. But not Passion and Pom — not even when Quis's threats became really vigorous. He displayed his huge canines more and more often, opening his mouth wider and wider in exaggerated yawns. He flashed the whites of his eyelids. He lunged towards the chimps, stiff-legged. He stood, 'champing' his jaws, making audible grinding sounds with his teeth. Above all he tried to catch the eyes of the chimps since, for a baboon, it seems difficult if not impossible to attack without first glaring his hostility directly into his opponent's eyes. To this end Quis circled round, directing his threats first at one chimpanzee and then another. Passion and Pom calmly ignored this — only young Prof, as was to be expected, was scared, repeatedly moving so that one of the females was between himself and the angry baboon.

Suddenly Pom, seemingly tired of licking, lay down. Quis, surely humiliated by her lack of respect, at once bowed right down and exposed his huge canines within inches of her face. But Pom, instead of shrinking from this close-up display of weaponry, reached up and playfully batted the irate baboon on the nose! Startled, he backed away, then yawned again — once more Pom, now with a wide play-face, hit out at him. As he continued to threaten her she sat, then stood, and began to hit at him more vigorously, still indicating, by her facial expression, that it was only a game. But Quis was unable to tolerate this insubordination. With an angry grunting roar he lunged at her and hit her on the head. At this Pom's playful mood broke: her hair bristled aggressively and she picked up a palm frond and flailed wildly at him. And Quis gave up. With all the dignity he could muster he stalked off and left the chimpanzees in control of the salt lick.

Sometimes a young chimp will tease an old male baboon in a most disrespectful way. Never shall I forget when Freud, aged five years, started to pester Heath of Camp troop. Heath was sitting peacefully in the shade, minding his own business, and seven chimpanzees were resting and grooming nearby. Freud climbed into the tree above Heath and began swinging over his head, kicking out at him playfully. For a while Heath showed remarkable patience. When Freud's foot poked into an eye or ear he simply turned his head away. But after ten minutes he had had enough. Leaping up he grabbed Freud, pulled him from the branch, and bit him. Freud began to scream at the top of his voice, although, in fact, Heath's teeth were old and worn and it is unlikely that he was hurting the infant much.

It was twelve-year-old Goblin, who had been lying some twenty-five feet away, who leapt to his feet and charged to Freud's rescue, cuffing Heath over the head. Freud escaped up the tree, Goblin returned to his rest, and Heath sat down again under the same branch. Peace was restored. But not for long. A few minutes later Freud, to my astonishment, began to tease the old baboon exactly as he had before. If anything he was even more irritating. Heath once more showed remarkable patience. But not Goblin. After a moment he got up and walked over towards Freud. With his hair slightly on end, and a furious expression on his face, Goblin reached up, pulled Freud down, and hit him hard. Freud, thoroughly disciplined and brought down to size, didn't even scream, but crept quietly away and went to sit near his mother. The old baboon settled down once again, minding his own business, in the late afternoon sun, while Goblin, still scowling, stalked back to continue his interrupted rest.

13

Goblin

‹‹‹‹‹‹‹‹‹‹‹‹‹

I FIRST SAW GOBLIN when he was just a few hours old in 1964. I wrote, at the time: '. . . Melissa, looking down, stared long at the tiny face. Never had we imagined such a funny twisted-up little face. It was comical in its ugliness, with large ears, small, rather pursed lips, and skin incredibly wrinkled and bluish black rather than pink. His eyes were screwed tight shut against the fading light of the sun and he looked like some wizened gnome or hobgoblin.'

Seventeen years later Goblin became the undisputed alpha male of his community. It was no easy victory that Goblin won as, for six stormy years, he challenged males older and, for the most part, considerably larger than himself. He risked much to acquire success, against odds that often seemed weighted heavily against him. His story is now an important part of Gombe's recorded history.

In retrospect I can see that, from an early age, Goblin showed many of the qualities that ensure eventual high dominance status in chimpanzee society. He was always determined to get his own way, he hated to be dominated, he was intelligent and courageous, and he could not tolerate disputes among his subordinates. The incident described at the end of the previous chapter, when Goblin first rescued, then disciplined Freud, is a typical example of his desire for social control.

In addition to these personality traits, a key factor in Goblin's early success was his extraordinary relationship with Figan, both before and during his time as alpha male. This had begun when Goblin was but a child. Without doubt it was Figan's presence, Figan's support, that gave Goblin the confidence to start challenging the other males at an unusually early age.

Like all highly motivated adolescent males, Goblin began to challenge the females of his community early and vigorously. In this endeavour Figan played little part, for these swaggering displays are seldom performed in the presence of adult males. Melissa sometimes helped her son on those not infrequent occasions when he fell victim to the retaliatory fury of one of the higher-ranking, tougher females. But she was not always around and Goblin frequently had to stand alone. As his displays became ever more vigorous and his confidence grew, he tackled the more senior females, and many were the times that he was chased off, often by two females who had formed a temporary alliance. Those incidents often ended in fights which, initially, Goblin usually lost. But even if he fled screaming he would be ready to tackle the same females all over again when next he encountered them. He never gave up.

It was during this period of his life that Goblin began to challenge me more and more often. From infancy onward Goblin, like Flint, had shown a tendency to 'pester' humans. When he was about four years old we realized that he was going to become a real nuisance. He would approach me, or one of the other students, and seize hold of our wrists. And there he would hang, gripping ever more tightly if we tried to shake him off. Note-taking became increasingly difficult when he was around. Eventually we hit upon the idea of arming ourselves with tins of grease — used engine oil, margarine, anything. When he approached us we quickly smeared our wrists and hands. And, because he hated to get his hands greasy, he soon learned to leave us alone. But as he moved through adolescence he began to pester humans in a different way — or, rather, he pestered *me*!

Chimpanzees can very clearly differentiate between human males and females. They are, by and large, far more respectful of men, particularly large men with deep, resonant voices. They take liberties with women. And I think Goblin seriously felt that it was necessary to dominate me along with the other females in his life. The fact that I was of a different species did not seem to worry him. And so I went through a trying few years, never quite knowing when Goblin might charge out of the undergrowth, run up behind me, and slap me or even stamp on my back. There were times when I was quite black and blue. This irritating — and sometimes painful — behaviour eased off after a while. I never reciprocated and so I suppose he reckoned that he had subdued me and I was no longer worth bothering about. Indeed, by the time he was twelve years old he was directing far less aggressive energy towards the female chimpanzees as well. As he had already physically attacked and thoroughly defeated most of them it would have been a waste of effort. He continued to challenge the remaining three — Passion, Fifi and Gigi. All three occasionally attacked him but Goblin took these setbacks in his stride: there would be another opportunity soon enough. When he was just thirteen years old he successfully conquered Gigi, the toughest of them all.

Now he was free to turn his attention to the lowest-ranking of the senior males, Humphrey. Poor Humphrey, fallen king, challenged by a youngster barely in his teens! At first when Goblin displayed towards him, Humphrey ignored him, or waved an arm in irritable threat. But Goblin persisted. At some point Humphrey must have realized that this was no ordinary thirteen-year-old show of valour: it signified the beginning of the end. Then Humphrey's irritation gave place to nervous tension and he began to respond in kind to Goblin's boisterous challenges.

This power struggle between Humphrey and Goblin clearly put Figan in an awkward position. His loyalties were divided between Humphrey, now established as his 'best friend', and young Goblin with whom, for so long, he had enjoyed a peaceful, almost paternal relationship. When he was present during one of their disputes, Figan typically compromised by displaying between the two, and this usually terminated the incident.

The first real conflicts we saw between Goblin and Humphrey took place at the end of 1977. Once, as Humphrey displayed towards him, Goblin whipped the older male with a sapling still rooted to the ground. Humphrey charged on and past him and Goblin started to feed. But not Humphrey. He sat glaring at the young male for another thirty minutes, as though brooding. And then he displayed at Goblin again. This time the two males stood upright and hit out at one an-

other, hair bristling. Humphrey started to scream, while Goblin remained quite silent. In the end it was Humphrey who lost his nerve
and, still screaming, left Goblin master of the field.

The second incident resulted in an even more clear-cut victory for
Goblin. Humphrey had just mated a pink female and was peacefully
grooming her, when Goblin approached, hair and penis erect, clearly
desirous of copulating in his turn. Humphrey at once charged furiously
towards his young rival. But Goblin, far from being intimidated, held
his ground. The two skirmished up in the branches, and Humphrey,
weighing about a hundred pounds compared to Goblin's seventy-five
or so, was actually knocked from the tree. He ran off screaming and
Goblin, after watching for a moment, returned to the female and
calmly mated her.

And so Goblin entered the hierarchy of the adult males when he
was only thirteen years old — at least two years earlier than other
males whose progress we have charted. Humphrey ranked below
him; five males ranked higher. In a variety of ways it was clear that
he was leaving adolescence behind. He spent more time grooming
with the other adult males, and sometimes they groomed him in return. He frequently joined in the charging displays that occurred when,
for instance, his group arrived at a new food source, or when two
groups met. He often mated pink females in full view of adult males,
rather than having to lure them to some private spot. After making
a kill, he was usually able to retain a reasonable portion, instead of
losing it all to his elders. And he began to take patrolling duties seriously.

All this time Goblin maintained his close relationship with Figan.
When the alpha displayed, then Goblin, if he was there, would join
in, following hot on his hero's heels, often imitating his actions. When
Figan performed one of his devastating early morning or late evening
arboreal displays, startling his screaming subordinates from their beds,
Goblin sometimes charged through the branches and swayed vegetation too.

The following year Goblin's progress was nothing short of spectacular. Systematically he began to challenge the senior males — first
the low-ranking, easy-going Jomeo, then Jomeo's kid brother, Sherry,

then Satan and finally even Evered. Only Figan was exempt. Indeed, it was his relationship with Figan that enabled him to challenge these older and more experienced males: he almost never did so unless Figan was nearby, and Figan, if he was there, almost always charged up in support of his young follower. Once, for example, Goblin and Evered began to fight each other when they were in a tree. Evered fought back and the two hung, hitting and kicking at each other, then fell to the ground. Goblin, clearly losing this particular fight, began to scream, at which point Figan charged up and Evered ran off.

Another incident took place when Figan was not around. It began when Goblin tried to move ahead of Satan when the group was travelling. This could not be tolerated and the much larger and heavier Satan attacked the younger male. Goblin ran off screaming, but an hour later, when Figan joined the group, Goblin at once began to threaten Satan, uttering waa-barks and displaying towards him. And Satan, no doubt anticipating the displeasure of his alpha should he retaliate, climbed hastily into a tree and sat there, whimpering softly to himself, as Goblin charged about below.

Soon after his fourteenth birthday Goblin, on a *one to one* basis, could intimidate all of the senior males — except, of course, Figan. And then came the day when, for the first time, Goblin was seen to challenge the brothers Jomeo and Sherry when they were *together*. Thrice he displayed past them as they groomed, going a little closer each time. And then, during a fourth challenge he actually hit Jomeo. Enraged, the brothers, each of whom weighed more than Goblin, chased after him. But though he ran off then, he did not give up. Four months later, almost exactly on Goblin's fifteenth birthday, there was a dramatic conflict. Jomeo and Sherry were grooming, and at first they ignored Goblin when he began displaying towards them — or at least they pretended to ignore him. But when he got really close they uttered fierce waa-barks and waved their arms. The situation grew increasingly tense, and when adult female Miff arrived on the scene she was immediately and violently attacked, first by Sherry and then Jomeo. In this way the brothers tried to vent some of their frustrated aggression.

Goblin made the most of this distraction. No sooner had Jomeo taken over the pounding of poor Miff than Goblin charged Sherry and attacked him fiercely. Quickly Jomeo left Miff and rushed over — but he helped his brother with vocal threats only. Goblin and Sherry fought on, rolling over and over, now with Goblin uppermost, now Sherry. They battled in silence until Goblin bit deeply into Sherry's neck and then, with loud screams, Sherry pulled away and ran off. Jomeo followed him, also screaming. And Goblin gave chase. For twenty yards or more he pursued them as they fled, then stopped and sat, gazing after them, his eyes hard and bright, his sides heaving. There were patches of saliva and fear-dung all over him. It was truly an amazing victory — and a decisive one. From then on, Goblin was able to dominate the brothers even when they were together.

It was in the following month that we saw the first sign of change in Goblin's relationship with his erstwhile hero. For some time we had been expecting Goblin to turn on Figan. Indeed, I am still puzzled as to why Figan, so socially adroit in all other ways, had not been able to predict the inevitable outcome of his sponsorship of Goblin. The first sign of disloyalty was recorded one peaceful afternoon when, instead of hurrying over to greet the newly arrived Figan, Goblin ignored him. After that, he ignored him more and more often and Figan, obviously sensing the implicit challenge, became increasingly tense and nervous. One day when Goblin suddenly appeared, Figan actually gave small cries of fear and ran, seeking reassurance, to embrace Evered. It became increasingly commonplace to see Figan, grinning in fear, running to seek help from one or other of the senior males. And from then on events slowly moved to their inevitable and foregone conclusion.

During the dry season of 1979 Figan somehow hurt the fingers of his right hand. He limped when he walked. Just as Figan himself had quickly seized on any sign of weakness in a superior, so did Goblin now. He began to challenge Figan in earnest, displaying towards him again and again, sometimes hitting him as he ran by. If one of the senior males was around, Figan always rushed to him for support. In this he was successful, and a strong feeling of unity grew between the five older males: they banded together, supporting the old order of things in the face of the young upstart. Thus Figan had four potential allies while Goblin, having alienated his only long-time supporter, stood on his own. He relied simply on the devastating effect of repeated vigorous and energetic displays.

Quite clearly, Goblin had profited from his close association with Figan — he had picked up a number of useful 'tips for dominance'. He had learned, for example, the psychological advantage gained by startling the other males from their sleep with a vigorous arboreal display above their nests early in the morning. And the value of surprise — hiding in the undergrowth when he heard a group approaching, then suddenly charging out. Both techniques gave results that must have been eminently satisfying to the ambitious young male. But it was clear that, for all his bravado, this was a stressful time for him. Again and again when confronted by pairs of senior males, Goblin revealed his tension by suddenly displaying towards females or youngsters apropos, apparently, of nothing. Once again I became a frequent scapegoat on such occasions. I remember once when Derek and I watched him as he tried to intimidate Satan and Evered, who were grooming. Again and again, seven times in all, Goblin charged past them, dragging branches and hurling rocks. Each time he went within a few yards of where they sat: they did not even look. Goblin became increasingly frustrated and after charging past the two males for the eighth time he carried on towards Derek and me. He avoided Derek, who was sitting beside me on the ground, then veered to give me a hearty shove with both hands and a double thump, thump with his feet before displaying away and then sitting, glowering at the world in general.

At the end of September we saw the first serious fight between Figan and Goblin. Goblin won, most decisively, kicking Figan from the tree into which he had fled. Figan fell some thirty feet to the ground and ran off screaming. A week later, after Goblin had displayed at and around him five times, Figan again took refuge in a tree. Never shall I forget that day when I sat and watched Figan, once the most powerful of Gombe's alphas, becoming more and more agitated and unhappy as the minutes went by. He moved restlessly, he scratched. Once, very cautiously, he began to climb down towards the ground, but Goblin, hair bristling, gazed up so ferociously that Figan, with squeaks of fear, retreated. I was vividly reminded of similar incidents when Figan subjected Evered to the same humiliation. On this occasion I had an interesting insight into Goblin's mood. Eventually he moved away from Figan's tree and joined Melissa who was sitting in the bushes nearby. He stretched out on the ground and she began to groom him. And then, almost imperceptibly, he reached for his mother's hand and began to play, gently, with her fingers. There he lay, relaxed and peaceful, idly tickling with Melissa. And when Figan, with utmost caution, crept down the tree and away, Goblin, though he followed the older male with his eyes, continued to play.

Quite clearly Figan could no longer be described as an alpha male. But neither could Goblin because, although he could boss each of the other males if he met them individually, for the most part he was still unable to control the situation when two or more of them were together. For a fifteen-year-old, Goblin's position was remarkable — but for this remarkable individual it was simply not good enough. It was clear that he would not rest until he had made it to the top and to this end he devoted himself tirelessly, displaying in the vicinity of the senior males at almost every opportunity.

Then, in mid-November, came the Great Attack, which, for almost a year, put Figan back on top. It started during a meat-eating session when tension often runs high and aggressive incidents frequently break out. Goblin, who had had no meat, displayed at Figan, who had. Figan, surrounded by potential allies, stood his ground. There was a dramatic fight as, for over a minute, the two males battled in a silence broken only by the sound of gnashing teeth. Suddenly, as though in response to some unheard clarion call, all the other adult males present — Evered, Satan, Jomeo and Humphrey — joined the fray, fighting under Figan's banner. With the odds at five to one, Goblin started to scream and struggled to escape. When finally he managed to pull free he fled, with Figan in hot pursuit and the other males charging back and forth, highly roused and screaming. Goblin was very badly wounded during the fight, sustaining a deep gash in his thigh which was still bleeding heavily an hour later.

After that, Figan regained some of his former confidence while Goblin, in his turn, was uneasy in the older male's presence. One month after the Great Attack, Figan had the satisfaction of seeing Goblin rush screaming from his displays. Better still, when Goblin took refuge up a tree Figan kept him there, tense and unhappy, for over twenty minutes while he sat calmly below. The tables had been turned. The other senior males, having gained in confidence as a result of the Great Attack, now supported each other even more enthusiastically against Goblin. Lesser males would have given up the struggle after a setback as serious as this. But Goblin, desperately unhappy with his present position, was made of sterner stuff.

Once his wounds were healed, Goblin, although for a while he avoided direct confrontations with Figan, once more began to challenge the other senior males. It was not long before the repeated displays, the constant disruption of social harmony again began to tell on them. Gradually, during the ten months following the Great Attack, Goblin regained his former position until, as before, he could dominate each of them when they were on their own. And then he set to work on the reinstated alpha. Poor Figan — his recently regained confidence, shaky at the best of times, wavered and broke. His best friend Humphrey had disappeared, perhaps a victim of the Kalande males. And, although Figan had then sought friendship with both Jomeo and Evered, and although they spent a good deal of time with him, he had no one whom he could really trust. When Goblin was around, Figan's appeals for help from the three remaining senior males became ever more desperate.

Within a few months Goblin had, once again, utterly intimidated the hero of his youth. Soon after this Figan himself disappeared. Perhaps he too fell victim to intercommunity aggression. Or perhaps he died, alone, of some disease. We shall never know. I grieved for his passing for I had known him for many years and long admired his intelligence and persistence.

With Figan gone, Goblin's disruptive displays became increasingly violent. And in response the senior males would sit close together and groom one another with almost frenzied concentration. The harder Goblin tried to disrupt their grooming, the more they groomed. And the more intently they groomed, the more reassurance they got from each other, and the longer they could ignore, or pretend to ignore, his tempestuous behaviour. Goblin became increasingly frustrated. For one thing it is much harder to threaten a rival if he will not flee or even meet your gaze. For another, his rivals were displaying overt signs of friendship and that was hard for Goblin to swallow. At all costs he must break up these grooming sessions.

But the older males, their eyes only a few inches from small patches of fur, were able to maintain a pretence of disinterest for over fifteen minutes. Again and again he displayed towards and past them. In between, panting, he sat and glared. Eventually, he passed the threshold of caution and became brave enough to actually attack one of the groomers.

These incidents were amazing to watch. One day, for example, Goblin suddenly arrived in the group that I had been following all morning, which included Satan and Jomeo. As soon as he appeared, the two seniors, as usual, approached and began to groom each other. Goblin stood, his hair bristling, and stared at them, but they paid no heed. After a few minutes he started one of his displays. The two senior males continued to groom, their concentration seeming almost fanatical. The females and youngsters screamed in a satisfying way and shot up trees. But Goblin was not interested in intimidating *them*, only his rivals. He paused, then displayed again, passing a little closer to the two males. They continued to groom, even more frenziedly. And so it went on.

Goblin displayed vigorously seven times until he had worked himself into a state of utter fury. Then, during an eighth display he attacked Satan, leaping into the tree above and stamping down onto the older male's head. Now the groomers were forced to respond. Screaming loudly they charged Goblin, waving their arms in rage. And despite the fact that his adversaries weighed 108 pounds and 103 pounds respectively, the eighty-pound Goblin stood his ground and tackled them both. For just over a minute they grappled and hit at each other — then, to my utter amazement, Satan and Jomeo fled while Goblin chased after them, hurling rocks. And then, as though to emphasize his statement, Goblin attacked Satan again. After that, as though the tension was altogether too much for him, Goblin left the group altogether.

On one occasion a similar confrontation, but this time with Satan and Evered, ended with no clear-cut victory for anyone. Goblin left the other two and, again, went off on his own. This time Hilali followed him. An hour later he encountered Fifi and immediately subjected her to a violent attack. Then, for good measure, he beat up Freud and Frodo as well. To the accompaniment of screams and waa-barks he displayed away and continued his solitary pursuits. Forty-five minutes after leaving Fifi, Goblin bumped into another female and she, too, was attacked fiercely and — at least so far as she was concerned — for no reason at all. In fact we can imagine Goblin still seething as he stumped through the forest, spoiling to vent his frustrated fury, really directed against Satan and Evered, on anyone he met.

There were many occasions when, during tense interactions between the senior males, Goblin would suddenly charge and attack some innocent bystander. Such scapegoats were usually adolescent males or females — including, of course, me. When anticipating one of Goblin's onslaughts I always stood up and held firmly onto a tree. Then, when Goblin pounded on me I was less likely to be pushed over — I have never much fancied the idea of a chimp pounding on my prostrate body. Usually Goblin would simply pound a few times on my back in passing. Three times his attacks were worse. Once he pulled me from my tree, deliberately pushed me to the ground, then kicked me. Another time he started to pull me after him down a slope and I was terrified of losing my footing and falling on top of him. Heaven knows what the consequences would have been. The third incident was, I think, the worst. It started with his usual tactics as he seized the little tree that I was clutching, leapt up and pounded my back hard with his feet. But then he swung round, leapt up facing me, and kicked my chest. As he did so his wide open mouth with its four gleaming and rapier sharp canines, was barely three inches from my face. Occasionally Goblin would pound on one of the field staff also, and I think we all, human and chimpanzee alike, hoped most fervently that he would resolve his dominance position to his complete and utter satisfaction as quickly as possible!

It was at about this time that Goblin began, quite systematically,

With eighteen-year-old Freud, Fifi's eldest son.
(Michael Nichols/Magnum)

Figan, alpha male 1973–1980.

Flint insisted on riding Flo even after Flame was born. Usually he climbed on her back, but sometimes he clung in the ventral position, like a small infant. Flame, whose hand and foot are just visible, is sandwiched between her mother and brother. (Hugo van Lawick)

Flint, unwilling to accept separation from his mother, pushes into the night nest with Flo and Flame. (Hugo van Lawick)

Pom fishing for termites.

Like all infant chimps,
Freud was depressed
during weaning. Here Fifi,
his mother, holds him
close to comfort him.

Wilkie, number-two male in the hierarchy in 1989, performs a charging display.
(Michael Nichols/Magnum)

Freud, his hair bristling, leads a reluctant Gremlin away on a consortship.

Goblin became alpha male in 1981, when he was only seventeen years old, taking over from Figan. (Ken Regan/HBO)

Left: Athena stole this bushbuck fawn from an adult male baboon. The discovery that chimpanzees sometimes eat meat and may hunt cooperatively and share the kill was one of the early revelations of the research at Gombe. (L. Goldman)

Below: Miff and Mel, six months before her death. As an orphan, Mel formed a surprising bond with the adolescent male Spindle and then was 'adopted' by Gigi.

A mixed party resting and feeding. (Hugo van Lawick)

David Greybeard grooms Hugo, who grooms Flo, who grooms Fifi. (Hugo van Lawick)

With Goblin, alpha male. (Ken Regan/HBO)

to terrorize Jomeo. Even though it was already clear that Jomeo was highly submissive to the younger male, Goblin lost no opportunity to charge and attack him — during reunions or other periods of social excitement. Indeed, Goblin persecuted him so fiercely that for a while Jomeo, unless he was with one of the other senior males, would leave the group he was with whenever he heard Goblin's distinctive pant-hoots nearby. And then, having reduced the Gombe heavyweight to a state of abject inferiority, Goblin began overtures of friendship. Suddenly he was grooming him more than he groomed any other male, sharing food with him, reassuring him in time of stress. The two became frequent travelling and feeding companions. In other words, they became friends — and Goblin, for the first time since turning on Figan five years earlier, had an ally. Not a very strong one, perhaps, but at least when he was with Jomeo, Goblin had a chance to relax and enjoy male companionship.

About a year after Figan's death the other males finally seemed to give up. Worn down by Goblin's repeated challenges, they let him have his way. And so, at seventeen years of age, Goblin became the undisputed alpha, able to control almost any social situation. Although he continued to display often, his performances were less violent and led less frequently to attack. At long last, things became more peaceful for the other members of his community.

Looking back over this fascinating story it is clear that, whether genetic or acquired, Goblin, like Mike, Goliath, and Figan before him, showed, in super-abundance, courage and persistence — the will to get to the top and to stay there despite setbacks. Can we point to any aspects of Melissa's early care that might have contributed to the development of these characteristics? She was an attentive and supportive mother, yet in no way over-indulgent. When Goblin got into difficulties during his early attempts to walk and climb, his mother usually left him to get himself out of trouble, even when he whimpered — unless he was really stuck in which case she quickly retrieved him. She was not restrictive but not overly permissive either. She was not a punitive mother, and was not always able to command instant obedience — Goblin learned early on that, if he went on trying, he could sometimes get his own way. Yet he was not spoilt — when it came to things that really mattered to her, such as weaning, Melissa imposed her will on her son. All in all she was, quite clearly, a good mother with respect to her child-raising techniques. And, to the extent that Goblin's behaviour was inherited, since she contributed fifty per cent of his genes, she was undoubtedly a good mother in this respect as well.

14

Jomeo

‹‹‹‹‹‹‹‹‹‹‹‹

JOMEO'S PERSONALITY was utterly different from that of Goblin. Where Goblin was fanatical in his determination to rise to a high social position and stay there, Jomeo, from adolescence onward, was almost entirely lacking in social ambition. He was the heaviest male we have known at Gombe, tipping the scale at just over one hundred and ten pounds, and he was a terrible enemy to individuals of neighbouring communities. Yet he did his best to avoid conflicts with the males of his own social group. A conundrum, was Jomeo, with a unique personality and a unique life history.

We know nothing of his childhood, for he was already a young adolescent when first I met him, in the early sixties. I seldom saw him in his family setting since his mother, Vodka, was shy and, together with her two younger offspring, Sherry and little Quantro, she spent most of her time in the southern part of the community range. Jomeo, however, became a regular camp visitor. In most respects he was a perfectly normal adolescent, but he did have one idiosyncrasy. When he came to camp with one or more of the big males Jomeo, like any other youngster, was seldom able to get a share of bananas. And so, like the other adolescent males, he quite often arrived by himself — which meant that we could hand him his very own bananas. This was when the odd behaviour showed itself — the moment he set eyes on the fruits he began to scream. Not just a few small screams of irrepressible excitement — which would have been quite understandable — but loudly, and for a couple of minutes at a time. Naturally, any chimps who happened to be nearby rushed to camp to see what was going on — and helped themselves to Jomeo's bananas. For at least six months he behaved in this peculiar fashion. And then, quite suddenly, the screaming stopped.

When he was about nine years old, Jomeo began his attempts to intimidate community females with the bristling, swaggering displays that are the hallmark of adolescence in the male chimpanzee. Initially, these performances were vigorous, impressive and audacious. Once he even dared to compete with Passion for a pile of bananas. As this most high-ranking and aggressive female began, with absolute confidence, to gather up the fruits, Jomeo stood upright and, with every hair on end, so that he looked twice his already large size, he swaggered in front of her, arms waving, with tight-lipped and furious mien. Passion, probably startled by his temerity (for to her he was still a child) gave back as good as she got and, as he displayed away, seemingly defeated, she began to gather up the scattered bananas. But Jomeo had merely gone to better equip himself for battle. Seizing a large dead branch that was lying nearby he charged back and began to swagger even more impressively, brandishing his weapon. And Passion, while she hung on to the bananas she had already picked up, did not dispute Jomeo's right to the rest.

It seemed then that Jomeo was firmly established on the ladder that would lead, ultimately, to a high position in the dominance hierarchy. But then something happened. One day in 1966, just a few months after his successful confrontation with Passion, Jomeo limped into camp covered with deep wounds. The worst was a great gash across the sole of his right foot which took weeks to heal and which left the toes permanently curled under. We shall never know who or what attacked Jomeo, but whatever it was that happened, it seemed to affect his whole subsequent career. His blustering displays towards the community females, even the lower-ranking ones, abruptly ended. A year later I observed an incident that typified Jomeo's position in his society. It began when Passion's infant Pom moved too close to Jomeo during feeding. When he hit out at her, warning her to keep her distance, she did not move but, looking towards her mother, then back at the big male, gave a small but defiant-sounding bark. Instantly Passion charged towards Jomeo — and this time, in marked contrast to his performance the year before, he fled before her and, screaming in fright, took refuge up a palm tree. When she began to climb after him, Jomeo, screaming even louder, leapt to another tree, tumbled to the ground, and raced, helter-skelter, away.

By that time, Jomeo had become the heaviest male at Gombe, and his chicken-hearted behaviour made him the laughing stock of his human observers. Even when he was fifteen years old and weighed nearly one hundred pounds, Passion could sometimes put him to screaming flight. And so it might have gone on, perhaps for the remainder of his life, had it not been for his brother, Sherry. The two of them had begun to spend more and more time together after the disappearance of their mother in 1967. Whether she had died, or perhaps decided to remain in some peripheral haunt, we do not know: she and her infant daughter simply stopped appearing in camp and were never seen again. But Sherry and Jomeo became all but inseparable, and in many ways the elder brother acted in loco parentis. When Sherry, during his early attempts to intimidate the females, was threatened — and like all young adolescent males he often was — then Jomeo ran to his defence just as Vodka would have done had she been there. As time went on and Sherry tackled higher- and higher-ranking females, so Jomeo's help was needed more and more often. And on those occasions when he did fight, Jomeo was a chimpanzee to be reckoned with. What matter if his technique was not always the best — he was still at least twenty pounds heavier than the largest of Sherry's female adversaries, and he inflicted hurt no matter *where* he hit or kicked. When he bodily lifted his victim in the air, then slammed her down, as he so often did, the punishment was horrible to watch. And so, at last, the females began to respect and even fear Jomeo and the days of Passion's supremacy over the huge male were gone for ever.

The frequency with which a male displays is, of course, an important factor in determining his position in the male hierarchy. Jomeo's frequency had dropped to almost nil after the horrible injury to his foot six years earlier. But now, because of his new self-confidence, he began to display much more often. Poor Jomeo — I sometimes wonder whether those early performances of his, intended to strike fear into the hearts of the beholders, were as amusing to the chimpanzee spectators as they were to us humans! He had so much to learn when it came to technique. Once, for example, he tried to enhance a fast downhill charge by rolling a huge rock. But instead of bounding noisily down the slope, adding a whole new dimension to Jomeo's performance, it remained firmly embedded in the hard ground. Any other male would have charged on regardless. Not Jomeo. He came to a complete halt, turned around, and heaved and pushed on the offending rock. Eventually he pried it from its resting place — but to no avail. It was much too big, and after rolling lazily for a couple of feet it came to a halt. Jomeo, the effect of his display now completely ruined, ran on, in a half-hearted way, without it.

Another time as he charged towards a group of females and youngsters, he tripped over a tree root and fell, sprawling in the undergrowth. The females, instead of screaming and fleeing in the way that must be so immensely satisfying to a young male, had quietly climbed nearby trees and, by the time he had picked himself up, were watching him from safety.

Funniest of all (from our point of view) was 'the case of the recalcitrant sapling'. It was a small tree, with a nice leafy top that would have looked good if flailed and brandished by a charging male. But when Jomeo seized hold of it as he ran past, he failed either to snap it off or to uproot it. And so, as with the boulder, he interrupted his performance to struggle with it. After about thirty seconds he finally managed to uproot the little tree. By then it was quite clear (to me) that it was far too large to make an effective prop. But Jomeo, having won his battle with it, was obviously determined to use it anyway. He charged on, dragging it tenaciously behind him. At least, that is what he tried to do. But it had so many side branches that one or other repeatedly became entangled in other vegetation: three times, before he finally abandoned his display, Jomeo was forced to move backwards, hauling at the sapling with both hands.

Gradually though, with the passing of the months, Jomeo's displays improved and he developed an impressive and powerful technique that was all his own.

It was the same when it came to hunting: at first Jomeo, though dead keen usually bungled the job. There was the time, for instance, when he tried to catch an adult blue monkey. The chase was fast and furious and the monkey, in desperation, took a flying leap to a neighbouring tree. Jomeo, close on its heels, likewise launched himself into space. But he never made it. 'Half-way across he simply ran out of jump,' David Bygott (who observed the incident) told me afterwards. Poor Jomeo: he crashed to the ground some thirty feet below, and for a chimp as heavy as Jomeo, that is some fall. He stayed quite still for a few moments, undoubtedly dazed and probably hurting. Then he stood up, gazed after his rapidly vanishing midday meal, and plodded off to eat figs.

When hunting, the Gombe chimpanzees mostly capture infant and juvenile prey, and typically they make heavy weather when it comes to killing an adult monkey. Thus when Jomeo captured a full-grown colobus male it was not surprising that it took him quite some while of hard and strenuous biting, flailing and hitting before his victim lay, limp and dead, across a branch. Then, before Jomeo could enjoy even a single mouthful of his hard-won prize, the other senior males converged on him and snatched it away. It was Richard Wrangham who watched that drama, and I remember him telling me the rest of the story afterwards:

'He sat and watched for a bit as the others divided up his prey. They were all excited and screaming and he was very quiet. He didn't join the females and youngsters to beg for a share. He just went and licked a few leaves below, where blood had splattered down. And then he wandered away. I almost felt like crying.'

As time went on there were other reports of Jomeo losing his prey to higher-ranking males — once even to Gigi — and we all began to feel sorry for him. But we noticed too that he very often disappeared during or after a hunt. And we began to wonder whether, perhaps, he sometimes managed to catch a small monkey during the confusion and sneak away with it before any of the others noticed. One day, after catching an infant which was then taken from him by Figan, Jomeo disappeared as usual. And then, about two hours later, he was found, sitting by himself, with a hugely distended stomach, and clutching the remains of a bushbuck fawn. Clearly it wasn't always necessary to feel sorry for Jomeo!

But all the time, despite his new accomplishments — his unchallenged authority over the females, his improved display techniques, and his increasing skill in hunting — Jomeo continued to be plagued by countless small indignities. All of which, of course, increasingly endeared him to us. There was, for instance, the day that I watched him as, slowly and with an air of intense concentration, he inched his way up a tall tree. All morning it had rained and the trunk, gleaming like polished ebony, was very, very slippery. At last the lowest branch, some twenty-five feet above the ground, was within the climber's reach — but even as he made a grab for it he began to slip. Faster and faster he plummeted earthwards, clutching the treacherous trunk tightly but in vain. There was a thud as the Gombe heavyweight reached the ground. For a few moments he sat perfectly still, staring at the trunk before him. Then, after gazing up into the branches above, he slowly rose to his feet and, with great determination, began the difficult ascent a second time. No fairground enthusiast ever struggled up a greased pole with more persistence — and this time he made it. For the next hour he feasted on tender green leaves and by the time he was ready to descend, the trunk had dried out in the afternoon sun and he reached the forest floor with dignity.

Then there was the colobus monkey incident. The adult male colobus are extremely brave in defence of their females and young. Even when chimpanzees are hunting in groups these colobus males will charge and mob them fearlessly and usually succeed in driving them away. Perhaps this is because, although the colobus are smaller, they have long sharp canines and they almost always try to bite the hunter's genitals. Thus it is not unusual to see two or more chimpanzees leaping away through the branches with loud screams, hotly pursued by a couple of infuriated colobus monkeys. But what happened to Jomeo one day was utterly unusual. He was sitting, peacefully feeding on fruit and minding his own business when a large male colobus assailed him — out of the blue as it were. Launching himself from a branch above, the monkey landed almost on top of Jomeo and hit at his head, uttering the curious high-pitched threatening call of his kind. Jomeo, surprised out of his wits, gave a single startled yell and fled!

'And who but Jomeo,' laughed Richard one evening, 'would be put to flight by the sight of three baby porcupines rustling noisily and busily through the dry grass!'

Even an event that was essentially tragic ended by making Jomeo more of a comic figure than ever. Somehow he hurt his left eye. For more than two weeks the lid remained tightly closed, quantities of fluid dripped out, and it was obviously very painful. We gave him antibiotics in bananas and eventually the wound healed, but it left him not only with impaired vision, but also with an eye half-white from scar tissue. He should have looked sinister — indeed, sometimes he did, especially when he peered out from thick foliage in the dim light of the forest. More often though it gave him a somehow rakish appearance. Poor Jomeo — not only the character but even the appearance of a clown.

Despite the fact that he eventually established his dominance over the adult females, Jomeo almost never showed any interest in bettering his position *vis à vis* the other males. He did have one long standing rivalry with Satan, who was about the same age as himself. We saw the first signs of this in 1971 when they were late adolescents and sometimes swaggered at each other with bristling hair when competing for food or during the excitement of a reunion. At that time it seemed that their social rankings were about equal, and these confrontations usually ended with the two rivals, grinning hugely, embracing one another. After a couple of years, though, Satan, after winning a few fights, asserted his dominance over the bigger male — unless Sherry was there to back his brother, in which case Satan, confronted by the fraternal team, would give way.

When Sherry began to challenge the lower-ranked of the senior males his displays were tempestuous, daring and imaginative. He would emerge suddenly and unexpectedly out of the bushes hurling huge rocks and flailing branches and fronds with such fiery zeal that the senior males would often get out of the way — thus boosting his ego so that he challenged his elders more and more often. Whenever his impetuosity got him into trouble, Jomeo — if he was there, and he almost always was — would charge over and display impressively in support of his younger brother. It seemed that Sherry was all set to rise to a high social position and there were many who predicted that, before too long, he would topple Figan — the reigning alpha at that time.

But then came a decisive defeat. Satan, exasperated by a long series of the younger male's disruptive displays, finally turned on him and attacked him fiercely, inflicting numerous wounds. Jomeo, as usual, rushed to Sherry's assistance and, although he did not actually attack Satan, displayed so violently around the conflict that Satan turned from his victim in order to chase the elder brother away. This almost certainly saved Sherry from even worse injuries.

That was a historic fight, for it brought to an end Sherry's bid for high social rank. After that, although he did sometimes fight the senior males, it was usually in the context of meat eating or sex — when, in other words, there was some immediate, material reward. But for the remaining few years of his life he never again strove for a high position for its own sake. Thus Sherry reacted to adversity rather as had brother Jomeo to that unseen attack ten years before. How different, these two brothers, from those males whose heroic struggles took them to the top and kept them there at whatever cost to themselves: Mike, Figan and Goblin.

But what of Jomeo's exploits with the fair — or should I say pink — sex? If a male can ensure adequate genetic representation in future generations, this will more than compensate for any apparent shortcomings in other spheres. Alas, in this respect too Jomeo was, by and large, a failure. It is even possible that he never fathered a single child. He lacked the nerve to compete aggressively with the other males in the excitable groups that surround popular pink females, he lacked the imagination to seize sudden opportunities for clandestine matings when his superiors were otherwise engaged, and he lacked the necessary social skills to persuade or bully desirable females to accompany him on romantic interludes *à deux*. Indeed, in this last respect, his record was dismal: he often tried to lead females away but usually he failed. To the best of our knowledge he only went on fifteen consortships in fifteen years and on almost all those occasions the females managed to escape from him before that crucial time during the last days of their swellings. Worst of all — oh poor Jomeo — seven of his ladies were, when he took them off, already pregnant with the offspring of other males.

Nevertheless, despite his idiosyncrasies and failures — or perhaps because of them — Jomeo eventually became a respected senior citizen in his community. He had so little interest in the power struggle of the high-ranking males that he posed no threat to those for whom status was of supreme importance. And so Jomeo was chosen as best friend, first by Figan (after Humphrey's death) and then by Goblin. And although both these dominance-oriented males had found it necessary to terrorize Jomeo and utterly subjugate him prior to accepting his friendship, once he had convinced them of his utter subservience he reaped the benefits conferred by alpha males on their acolytes — some protection from other senior males, and a certain degree of tolerance in feeding and sexual contexts.

Jomeo came to represent security for the young males, too. Often, during their early journeyings away from their mothers, it was old Jomeo to whom they turned for companionship, sensing his benign tolerance. Once I followed as he wandered from one food patch to another with no less than five adolescent males trailing peacefully in his wake. During the five hours that I was with them I did not see him threaten any of them — not even when they fed very close beside him. There was one time when Jomeo stood upright and reached high above him for a length of succulent vine that he spied coiling around a branch. No sooner had he pulled it down and commenced to chew on one end than Beethoven approached, took hold of a piece of it — where it bifurcated at the end — and began to chew as well. Granted Beethoven was his favourite, but even so I was amazed that the big male made not the slightest gesture of protest.

I have wondered so often about Jomeo's fascinating character, his strange lack of any sort of dominance drive. If he had not been wounded as an adolescent, would he have gone on to become a high-ranking male? Probably not for, after all, his brother Sherry showed the same inability to cope with adversity. Was this a genetic, inherited trait? While this is possible, I suppose, it seems far more likely that it stemmed from the personality, the child-raising techniques, of their mother, Vodka. It is indeed unfortunate that I did not know Vodka well — she was too shy. But so far as we could tell she was a very asocial female, spending most of her time wandering, with her family only, in peripheral parts of the range. Prof, son of asocial Passion, has never shown any sign of wanting to dominate his fellows either. On the other hand, Figan and Goblin, who rose to be top-ranked males and who never accepted defeat for long, had mothers who were not only dominant, but also highly social — Flo and Melissa.

15

Melissa

❮❮❮❮❮❮❮❮❮❮❮

Melissa clearly merits special attention, if only as mother of one of Gombe's most dynamic alpha males. Her life was remarkable in other ways too. For one thing, she gave birth, in 1977, to Gombe's only known twins. I shall never forget my first sight of the babies — fraternal twins whom we named Gyre and Gimble. Melissa was sitting in the late afternoon sun, holding the two tiny bodies close to her breast so that they were all but invisible. One was suckling, the other seemed to be sleeping. When Melissa moved off, followed by her daughter Gremlin, I went with them and by the time I got back to the house that evening I had a real appreciation of the enormity of Melissa's task. Most infants, by the time they are two or three weeks old, can cling to the mother without support for long periods of time. The twins gripped well enough. But one kept clinging to the other by mistake: then he would pull his brother loose and both would start to fall, uttering loud cries of distress. Melissa had to give them almost constant support, holding them close with one arm, or travelling with her legs bent so as to support their backs with her thighs. Once, that first afternoon, one of the twins half-fell and banged his head on the ground. He screamed loudly and this set the other crying too and it was several minutes before Melissa managed to quieten them. She also had a lot of trouble making her nest. I couldn't see her well, for she was in dense foliage, but I heard a lot of crying from the babies.

That evening Derek and I talked with Hilali, Eslom and Hamisi around the fire. Hamisi described his first observation, when the babies were only a few days old. Melissa had travelled very slowly, walking a few yards at a time, then sitting and cradling the twins for a minute or two before moving on again. She had seemed exhausted and had made her nest early. The following morning Eslom managed to climb high in a neighbouring tree so that he could see into the nest. Gremlin left her small bed just after 7.00 a.m. and began feeding nearby. But Melissa showed no signs of stirring for another hour and a half. Then she sat up and began to groom herself, occasionally making a few grooming movements on one or other of the twins. Ten minutes later she stood up, ready to go: the infants at once began to whimper. So Melissa sat down again, looked helplessly at the babies for a moment, and then lay down. Fifteen minutes later she again tried to leave — as before both babies began to cry and so Melissa, after cradling and briefly grooming them, lay down once more. The same thing happened several more times, and it was almost two hours after her first attempt to leave that she finally made it. Clutching the twins tightly and ig-noring their frenzied screaming, she climbed from the tree with almost desperate haste. Only when they were all three safely on the ground did she stop to comfort them.

For the first three months of the twins' lives Melissa was followed every day for we all feared that Passion and Pom would strike again and we planned to intervene if they did. And in Melissa's mind too the memory of their bitter attack on her previous infant must have been vivid for, despite the difficulty she experienced in travelling with her two babies, she managed, during the first month, to keep close to one or other of the big males almost all the time. The advantages of this were made very clear to me one day when the twins were about a month old. I had followed Melissa, Gremlin and Satan as they climbed to the very top of the mountain ridge we call Sleeping Buffalo. It was a grey, cold afternoon in November, with thunder grumbling in the south. Earlier it had rained hard, and our high valley was still dank and chill under the sullen clouds. I shivered as I watched Melissa feeding on palm nuts above me. Suddenly a twig snapped: I turned sharply and found, to my horror, that Passion and Pom had ap-proached, moving almost without sound on the soft, wet, forest floor. Now they stood, motionless, staring up towards Melissa and her babies. None of the chimps above had seen them. With slow, stealthy movements Pom started to climb towards Melissa. Passion, heavily pregnant, climbed as well, but she soon stopped and watched from a low branch. Pom, creeping very quietly, got closer and closer and I was just about to yell a warning when suddenly Melissa saw them. Instantly she began to scream, loudly and urgently and, reckless in her panic, took a huge leap through space, towards the nearest branch of the next tree, the babies supported only by her thighs. My heart was thudding but somehow all three made it and Melissa hurried to sit close to Satan — who had stopped feeding and was watching Pom intently. Melissa, with one hand laid on the big male's shoulders, turned and barked her defiance at the younger female. And so the attempt was foiled. But if Satan had not been there it seems almost certain that another gruesome battle would have taken place, high above the ground, and there would have been nothing I could have done to help.

Very soon after that incident the twins developed quite bad rashes on their bellies and inner thighs, and Melissa, we noticed, had lost a good deal of hair in the region of her groin. This was because all three were becoming fouled with urine and faeces. Usually a baby's excrement falls neatly between the thighs of the mother as she sits — and if, by chance, there is a mistake, the mother quickly picks a handful of leaves and wipes herself clean. But it did not work this way with the twins, and Melissa simply could not cope. And then, on top of this, Gyre somehow hurt his foot. Clearly he was in considerable pain, for he screamed loudly almost every time Melissa moved — a strange scream, high-pitched like the wild cries of some sea bird in distress. Poor Melissa — the crying of one sick twin was bad enough, but so often Gimble joined in, frightened, perhaps, by the intensity of his brother's calls. Sometimes when they yelled Melissa sat and cradled them until they quietened. But at other times, holding them tightly, she moved on very fast, uttering a series of cough-like grunts — as though she was threatening them. Usually they then screamed even louder and, after a few minutes Melissa, utterly confused or fed-up (or both), climbed into a tree and, with the same quick movements, constructed a large nest. During this process the screams were redoubled and could be heard from afar. But once Melissa lay in her nest with them, all was quiet.

Now that Melissa could no longer keep up with the big males, she and Gremlin spent much time in the vicinity of camp. It was fortunate indeed that Passion, heavy with child, was no longer interested in devouring the infants of others. And Pom, although she could almost certainly have snatched one of the twins without much difficulty, clearly lacked the nerve to tackle an older female without the support of her high-ranking mother. However, while the danger of a cannibalistic attack thus seemed remote, we were worried on another score. For Melissa, preoccupied with the task of transporting and quietening her twins, spent less and less time in feeding. Some days, indeed, she fed for only an hour — whereas normally an adult chimpanzee spends some six to eight hours a day in feeding. We gave Melissa extra bananas, and the men gathered some of the wild fruits that were in season and offered those as well.

After a week I decided that we should give Melissa a course of antibiotics. I hoped that this would get into her milk and help to clear up Gyre's infected foot. And so, for five days, we took a little supply of bananas with us when we followed Melissa and, at regular intervals, handed her one — laced with medicinal powder. I don't know if this helped, but Gyre's foot did get better and soon Melissa was able to go about her daily business with no more difficulty than before.

Gyre's injury, however, was a setback from which he never really recovered and from then on it was clear that Gimble was developing much faster than his twin — and even Gimble was way behind a normal youngster. It was not until he was six months old, when most infants are already taking their first steps, that Gimble began to pull himself to different positions on his mother's body. Yet once he began these exercises, Gimble was soon able to scramble up onto Melissa's back. Having mastered this skill he very often rode on his mother during travel, or draped himself with his head hanging over her shoulder as she sat feeding. Sometimes he even slept like that. Probably he liked getting away from the somewhat crowded accommodation offered by the maternal lap. It was not until he was ten months old that, for the first time, he broke contact with Melissa to take his first tottering steps and climb onto his first small branch. Gyre, however, never even tried to walk or climb. He remained quietly on his mother's lap, often with his eyes closed.

The 1978 dry season was unusually harsh and by August there was less food than usual at Gombe. Even before this there had never seemed to be quite enough milk for two infants, and now it was obvious that both twins were permanently hungry — at almost any minute of the day there was one, if not two, infants tugging urgently at Melissa's breasts. Gimble, stronger and more active than his brother, almost certainly took more than his fair share of the scant supply, and so Gyre became more and more lethargic. When he fell sick with the cold that was going the rounds, his weakened system could not cope. The cold became pneumonia and one day Melissa arrived in camp carrying Gyre, a small limp body, in one hand. He was too weak to cling to her, he breathed with difficulty and his eyes remained closed. When Melissa climbed into a tree, supporting Gyre only with her thighs, he fell, landing with a thud on the hard ground some ten feet below. Melissa rushed down to gather him up, embrace him and groom him. He was still breathing when she moved off, but she carried him as though he were already dead, slung over her shoulder and held in place with her chin. He fell several times, lying motionless on the ground until she picked him up. The following morning he was dead.

I was sad when Gyre died, and disappointed at the lost opportunity to chart the development of twins in the wild and study the relationship between them. Nevertheless, I couldn't help but feel that, for both Melissa and Gimble, it was for the best. Certainly Gimble then began to make up for lost time. Soon, although tiny for his age, he was practising acrobatics in the branches and playing with other young-sters. He became increasingly active, gambolling from place to place, performing little stamping displays, turning somersaults, and, on many occasions, playing wildly with fallen leaves. Sometimes he swept them into a great pile with his hands, then moved backwards, pulling them after him. Or he pushed them ahead of him in a mound that grew bigger and bigger as he went. Often he rolled in the leaves, and once he threw handfuls all over his head and back, then rubbed them against his face.

Melissa still had her problems, but they were different now. Gimble often refused to follow her when she was ready to leave: either she had to drag him away, or she had to wait. Once, as she tried to pull him after her, he seized hold of vegetation with both hands and clung on for some moments before she could wrench him free. Eventually she got him onto her back, but after she had taken only a few steps he leapt down and ran back to play. Quickly Melissa grabbed him again and dragged him after her. Soon he escaped and once more ran to play. Melissa chased after him, but he avoided her and hid behind a tree. Melissa followed and, as he gambolled away, grabbed at him — and missed. He began to play again. Melissa watched for a moment, then reached out cautiously, seized his hand and began to pull him over the ground after her. Gimble bit her hand but it was only in fun and she reciprocated, tickling him. Soon he was laughing loudly. After that she once more put him on her back and this time he stayed on board.

Throughout Gimble's infancy Gremlin was an integral part of the family. At Gombe there is no closer relationship in chimpanzee society than that between a mother and her grown daughter. Females seldom start to leave their mothers, even for a few hours, until they are about ten years old — and then only when they are sexually attractive. There are definite benefits for a young female who hangs around with mother. For one thing, she can often get the better of females older than herself because her mother usually intervenes on her behalf if things go wrong. The mother typically joins forces with her daughter against the early challenges of young males, too. But it is not all roses. The young female has to pay a price for this protection and support: her mother will dominate her utterly, showing all the authoritarian discipline of a Victorian matriarch. Thus it is Mother who chooses the direction of travel, Mother who decides whether they should move fast or slowly, Mother who automatically gets first choice of feeding place and food. Gremlin, like all the other young females, soon found this out for herself.

When, for example, they were termite fishing Melissa repeatedly displaced Gremlin from her working place, or reached over and took her daughter's tool. At first Gremlin often threw tantrums. I remember one occasion when Melissa seized a spendid long tool that Gremlin had just fetched and prepared: Gremlin hung on tight, whimpered, then gave little screams. At this Melissa embraced her until she quietened — and *then* took the tool! But as time went on Gremlin became increasingly philosophical: she sometimes whimpered a little when her mother robbed her thus, but then she would move off to find herself a new place to work, or to pick herself another tool. Sometimes Melissa only needed to look towards her daughter with, presumably, an acquisitive gleam in her eye, for Gremlin to relinquish her claim to a food item — to a tunnel in a termite mound or a fruit-laden branch, for example. When Gremlin got to a tree ahead of her mother, and she determined, after gazing up, that there was only a limited supply of food, she would often move away of her own accord, leaving the field clear for Melissa. This was as it should be. Melissa had suckled Gremlin and shared food with her for years — now it was important that she take over the richest supply so that she could build up her strength and nurture other youngsters. And Gremlin, with only her own healthy body to care for, not only had lower nutritional requirements, but also the unbounded energy of youth. Moreover, she could feed high among the slender branches that were out of bounds for her heavier mother.

Of course, Gremlin was quite free to leave her autocratic parent whenever she liked — but then she was at the mercy of all the females who, when she was with her mother, showed her respect. Moreover Melissa, for all her selfishness over matters connected with food, was immensely supportive of her daughter in other ways. Most dramatic was the time when Satan attacked Gremlin and, in response to her daughter's screams, Melissa actually leapt at the big male, hitting and biting him. She got very badly beaten up for this interference. And so Gremlin, like most daughters, chose to remain firmly attached to the maternal apron strings.

There is no question but that the mother–daughter bond is highly beneficial for the mother, too. Gremlin was loyal and valiant in her defence of Melissa. Once, when still a child, she had even tried to rescue Melissa from a brutal attack by Satan. The fact that she was much too small and light to be of any real help does not diminish her valour. She had hurled herself at the big male, hitting him with her fists, then run over to Goblin who was nearby and pulled at his hand, while looking repeatedly from him towards their embattled mother. Clearly she had been begging him to help. But Goblin, whose relationship with Satan at the time was very tense, had been in no mood for chivalry and he had just sat and watched. So Gremlin once again had hurled her puny self, courageously if futilely, into the fray, joining Melissa in uttering loud barks of defiance at Satan when, finally, he charged away.

She had behaved in the same valorous manner when Melissa had tried to save infant Genie from Passion and Pom. Time and again Gremlin had leapt at the murderous females, beating at them with her small fists. She had even run over to the field staff for help. Standing upright in front of them, she had looked into their eyes, then turned to where Melissa battled for the life of her infant, then back towards the men. They had known she wanted them to help, and they had wanted to intervene; but the battle had been too fast and furious. Feeling helpless, they had done nothing. So Gremlin had run back on her own and had hurled herself at her mother's assailants just as Pom had pulled the baby away from Melissa. And her intervention had been so fierce that, just for a moment, Melissa had actually managed to retrieve her infant — only to have her once more snatched away. For good.

As Gimble got older Gremlin became increasingly helpful to her mother in one other way — she shared in caring for her young brother. If only Melissa had let Gremlin help when both twins had been alive, how much easier her task would have been. Instead, confused by the burden of caring for two babies, she had been unusually protective, and had forced Gremlin to keep her distance. By the time Gimble was three years old, however, there was scarcely a follow when Gremlin was not seen carrying him for part of the time; and when the family was feeding peacefully, Gimble was frequently closer to his sister than to his mother. If he got into any kind of trouble it was often Gremlin who responded to his whimper or scream of distress, running to gather him close. Once adolescent Atlas, when mating Gremlin, had hit out angrily as Gimble rushed up to push between the couple. Gremlin, outraged, terminated the copulation abruptly, turned and attacked Atlas.

Gremlin's concern for Gimble went way beyond merely responding to his appeals for help: like a good mother she would anticipate trouble. Thus when Gimble played with young baboons Gremlin often watched closely and, if the game got the least bit rough, and long before Gimble himself seemed worried, she firmly took him away. Once, as she was carrying him along a trail, she saw a small snake ahead. Carefully she pushed Gimble off her back and kept him behind her as she shook branches at the snake until it glided away. Another time Gremlin, with Gimble perched on her back as usual, suddenly stopped just before the trail passed through a patch of tall grasses. Melissa carried on but when Gimble, who had jumped to the ground, tried to follow his mother Gremlin prevented him. She pushed him behind her, hit a few times at the grass ahead, and then herded him around the clump of grass. I expected to find another snake hiding there — instead I saw that it was infested with hundreds of minute ticks.

Gremlin was very tolerant of her small brother. During the termite fishing season, an infant will often seize the opportunity to poke into a hole that has been temporarily vacated by a chimpanzee searching for a new tool. Usually the child will be gently but firmly pushed away as soon as the rightful owner returns, but Gremlin sometimes sat for five minutes or more watching her young brother as he experimented with various abandoned tools, only reclaiming her hole when he gave up. Once, when he was a bit older, Gimble tried to take over the hole when his sister was still working it and when she prevented this he had the audacity to threaten her, raising his arm and giving a childish waa-bark. Gremlin paid no heed to this mixture of disrespect and sheer cheek, just gently pushed him aside and went on with her work.

No wonder she was a good mother to her own first infant, Getty, efficient and assured in her handling of him right from the start. A truly wonderful relationship developed between Getty and his grandmother. Melissa first set eyes on Getty when he was one day old — she had not been present during the birth, for Gremlin like most females had gone off on her own. When Melissa approached, that first time, Gremlin backed away nervously fearing, perhaps, that her domineering mother would try to appropriate this new and precious possession just as she took everything else. But Melissa sat quietly nearby and merely glanced at the new infant from time to time, and soon Gremlin relaxed. Not until Getty was ten months old did we see Melissa touch her grandson at all — and then it was merely to groom him for a few moments during a session with Gremlin.

Soon after that I watched a fascinating incident. It began as Melissa was grooming Gremlin's back and Getty pushed his way between them. Melissa looked down at him, then lifted him into her lap and began to groom him — just as though he were her own infant. Gremlin glanced round, then seemed to stiffen. Very slowly she turned; very cautiously, glancing into her mother's face, she reached towards Getty with a soft pleading whimper. He responded at once and climbed into her arms. Quickly Gremlin moved away, settling to rest some five yards distant. Clearly, once again, she had feared that Melissa might try to steal her beloved son.

As the days went by, Melissa seemed to become more and more enchanted by Getty and the bond between them grew. When Melissa and Gremlin were grooming together Getty repeatedly interrupted, leaping down onto his grandmother from some overhanging branch — and Melissa, who had never played much with any of her own offspring, would stop grooming and start to tickle him. During these games, which sometimes lasted for fifteen minutes, Gremlin usually sat watching. Melissa actually initiated some of the play herself — sometimes she even followed Getty when he was romping with another youngster and pulled him away so that she could play with him herself. This was not always to his liking, for he was a self-willed little fellow, and then he would struggle until he had escaped from Granny and could run back to his chosen playmates.

Of all the infants I had known at Gombe, Getty was the most endearing. He was lively and adventurous, always eager to join in any social activity. He was well able to amuse himself, too. Once, as Gremlin fished for termites, Getty played with sand for over ten minutes. He lay on his back with his mouth wide open, scooped up handfuls of loose soil and, holding his hands up in the air, dropped the sand so that it showered all over his body and into his mouth.

When Gimble was six years old Melissa resumed her sexual cycles. This led to the most extraordinary series of incidents; Goblin, who was now nineteen years old, suddenly evinced an incestuous sexual interest in his mother. During Melissa's previous pinknesses Goblin, like other mature sons, had shown absolutely no desire to mate with his mother. But this time it was different. One day, about half-way through her first period of swelling, Goblin approached Melissa and summoned her with vigorous shaking of vegetation. She ignored him at first and then, when he persisted, she threatened him. This seemed to enrage him — with a scowl he leaped at her and, as she ran off, chased after her and actually stamped on her back. Melissa was beside herself with fury and, as Goblin displayed away, she stamped after him, screaming until I thought she would choke. He left then, but the following day he summoned her again and, when she tried to avoid him, once more threatened her with bristling hair. Then, to my utter astonishment, Melissa actually crouched before her son for copulation. The sexual act was not completed — Melissa pulled away, screeching loudly, after a few seconds. Again Goblin leapt at her and stamped on her back. His own mother! I couldn't help but feel incensed — and clearly Melissa felt much the same for she turned and hit him before running off. She climbed high into a tree, as far away from Goblin as she could get. He stayed below, glaring up and angrily shaking branches, but she stayed put and soon he gave up.

After that we followed her every day until her swelling was gone. Goblin made a few more half-hearted attempts, but we did not see any further violence between the two. Nor was he aggressive towards her during her next pinkness, a month later: he did attempt to copulate with her a couple of times but she managed to escape — unviolated.

Goblin's unnatural behaviour utterly changed the relationship between Melissa and her son. They had been close before, spending much time in one another's company as they fed or travelled or rested. They were frequent grooming partners, too. Often Goblin hurried to help his mother, whether in her dominance interactions with other females, or when she was being challenged by some callow adolescent male. After Goblin's attempts to mate his mother, however, relations between them were very strained and tense. Not only did they stop spending time together, but Melissa actually seemed to be frightened of her son. However, during her second period of swelling she became pregnant, after which, like most of the older females, she showed no further periods of oestrus. And so things between Melissa and her son slowly returned to normal. Moreover, even before that, during the height of their temporary estrangement, I saw something which showed that deep down their old relationship was still alive.

It happened when there was a high level of excitement among the chimps, for six females in addition to Melissa were cycling, flaunting their provocative pink bottoms. All the males were there, and most of the other community members, too. They travelled in noisy, boisterous groups, calling back and forth to each other across the valleys. A carnival atmosphere prevailed. The adult males displayed magnificently, the juveniles and infants romped and wrestled and chased one another through the trees. There were sudden outbursts of screaming as the excitement boiled over and led to aggression. Just occasionally, though surprisingly seldom, there was a serious fight. One of these took place in a tree right over my head — and the victim was Melissa. She was sitting quietly on a branch grooming young Gimble when Evered, who had been threatened by Satan when he courted one of the other females, suddenly leapt at her. She screamed and tried to escape, and as she did so I saw his teeth slash at her pink swollen bottom so that watery blood poured down. At that moment I heard a crashing behind me and Goblin hurtled past me and up into the tree. Without pausing he attacked Evered. All three were locked in combat no more than six feet above my head. I dared not move for the slope was steep and rocky and I was balanced against the trunk of the self-same tree, so I stayed where I was, praying that the branch would not break and deposit its enraged and screaming burden on top of me. Fortunately the fight ended, as it had begun, up in the tree — except that Evered leapt to the ground, and fled, screaming. Goblin stayed for a little while and watched as Melissa picked leaves with which to dab at her bleeding bottom. And then, as peace returned, he too climbed down and moved away.

The following day Melissa's swelling was shrivelled — a typical response to physical injury — and she was no longer interesting to the highest-ranking males. But she was to Jomeo. I met the two of them, with Gimble trailing along, quite by chance in Kasakela Valley. Poor Melissa — her bottom was sore and hurting and, on top of that, she had terrible diarrhoea and kept crouching forward as though suffering severe stomach cramps. And instead of being free to recover in peace she was being forced, by Jomeo, to follow him northward. A less likely-looking honeymoon pair would be hard to imagine, for Jomeo was in even worse shape than Melissa. The entire left side of his face was hugely swollen from jaw to eye and the flesh was an ugly shade of pink beneath the tight-stretched skin. He looked, with his one half-white eye, almost grotesque. To complete the pathetic picture, Gimble was in the midst of his weaning depression. He was keeping close to his mother with a sullen expression on his face, his lips pushed forward almost continually in a disgruntled-looking pout.

When I arrived the three were sitting, Melissa and Gimble close together, Jomeo a few yards ahead. He must have had an abscess on one of his upper molars and I think it burst right then, as I watched, because suddenly he began to dab his gum with one finger. He licked the finger, dabbed and licked — on and on. Gimble was fascinated, and peered closely as the big male tended his sore mouth.

Presently Jomeo got to his feet, moved a few yards further from Melissa, looked back and shook some branches. Melissa ignored this summons completely. Then Jomeo began to sway and swagger until every hair stood on end, and I felt sure that Melissa would be attacked. But at the last moment she obeyed and hastened to him with submissive pant-grunts, bowing to kiss his thigh as he groomed her. Ten minutes later Jomeo set off again, and the whole performance was repeated until, reluctantly, Melissa moved on another few yards.

I followed them for most of the rest of the day. We didn't go far — Melissa saw to that. In between Jomeo's efforts to move on, the three of them sometimes fed, but often they just sat. Jomeo dabbed at his gum. Melissa crouched or huddled, as though in pain, and, from time to time, plucked leaves to dab at her wounded bottom. Gimble repeatedly pestered his mother, demanding access to her nipples. When he approached her with his pouting face, whimpering and crying, Melissa was too weary and sick to protest for long. She gave in and he crept into her arms and suckled. When I left them, Melissa was lying with her eyes closed, one arm over Gimble who held a nipple firmly in his mouth. Jomeo waited nearby, dabbing at his abscess.

That consortship, like most others in Jomeo's life, was not successful: two days later the little trio reappeared in the central part of the Kasakela range. And the following month Melissa went on a consortship with Satan — and conceived.

About two months before we reckoned that Satan's baby was due, Melissa became very sick indeed. Her symptoms — bad cough, heavy mucous discharge and high fever — suggested pneumonia, and we feared for her life. She was unable to climb trees for several days and, at her worst, could barely drag herself along the ground. She ate only a few mouthfuls of food, refusing offerings from the concerned field staff. Amazingly she recovered, although her vocal chords were permanently impaired and her voice, for the rest of her life, came out as a hoarse croak. And, before she was properly better, her pregnancy ended with a miscarriage.

But then, three months later, Melissa once again travelled the hills flaunting the pink sex signal of the female chimpanzee. Almost at once she became pregnant — for the last time. How much better it would have been had she not. That last pregnancy sapped her strength and vitality and when little Groucho was born, Melissa looked frail and much older than her estimated thirty-five years. From the start Groucho was tiny and lethargic. When he was nine months old he occasionally made short forays from Melissa's side, began to eat solids and, occasionally, played gently with Gimble, but then his condition worsened. By the time he was one year old he was spending most of his time lying listlessly on his mother's lap. Gimble still tried occasionally to persuade this small brother to play, but Groucho, although he usually responded with a play-face, was too weak for the rough and tumble games typical of his age.

It was at this time, when I was almost expecting to hear that Groucho had died, that I received news — a telephone call from Kigoma — that Getty was missing. I shall never forget the sense of shock and outrage I felt when I arrived at Gombe a week later, and heard that his body, when it was eventually found in the forest, had been horribly mutilated — the head had been cut off and removed. We never discovered exactly what had happened, but we suspected witchcraft, for the old customs are deeply entrenched among the Waha people of the area. Nothing like this had happened before — nor has it happened since. It was a bitter blow for Getty had been the favourite youngster of us all. I feel quite sure, too, that among the chimps it was not only the members of his immediate family who missed him. Getty, with his adventurous and fun-loving nature, had captivated us all.

Gremlin was listless for weeks but eventually, two months after losing her son, she once more resumed her sexual cycles. Then she began spending more time with the males and less with her old mother. Gimble quite often left Melissa too. Goblin, however, now that his relationship with his old mother was back on course, travelled with her periodically, though never for long at a time. One day as I followed them through the forest we heard the pant-hoots of Satan and Evered across the valley. Despite his alpha rank, Goblin's relationship with the much heavier Satan was often tense. He stared towards the calls, his hair bristling, then turned to his old mother and, with a grin of fear on his face, stretched his hand towards her. She responded at once, reaching to touch his fingers and Goblin was calmed, just as he had been calmed throughout his infancy, by the contact with her. He turned and moved on to face whatever challenge lay ahead. Melissa followed for a while but soon she stopped to rest.

A few months later, as I was walking along the Kakombe Valley, I saw Gimble carrying something large into a tree. It was the dead body of little Groucho. As Melissa and Gremlin groomed each other on the ground, Gimble cradled the corpse on his lap, grooming it intently. When his family moved on Gimble climbed down and followed, the body slung over his shoulder. Presently it fell to the ground, and then he dragged it behind him by one arm. Later, when they stopped to rest again, Melissa gently took the limp body from him and placed it over her own back. She carried the dead baby for two more days and then abandoned his corpse deep in the forest.

After the death of her infant, Melissa seemed to lose the will to live. She had been thin before, now she became emaciated for she ate almost nothing. Often she did not leave her nest until after ten in the morning and sometimes she went to bed as early as four o'clock. During the hours in between she made at least one day nest where she lay, often staring vacantly up through the leaves, for hours at a time. Sometimes Gimble was with her, but he became bored, as well as hungry, and spent more time than before with the big males. Nor was Gremlin there to provide comfort: protesting, she had been led off on a two-week consortship, by Satan, on the evening of the day Groucho died.

Ten days after losing her infant, Melissa, using the last of her strength, climbed high into a tall, leafy *mgwiza* tree and there, surrounded by clusters of purple, sloe-like fruits, she made a large nest — the last she would ever make. Throughout the following day she lay, scarcely moving, while other chimpanzees, attracted by the succulent fruits, arrived, fed for an hour or so, and left. Gimble was nearby for much of the day, and sometimes groomed his mother. But he moved away during the afternoon.

By evening, Melissa was alone. One foot hung down from her nest and every so often her toes moved. I stayed there, sitting on the forest floor below the dying female. Occasionally I spoke. I don't know if she knew I was there or, if she did, whether it made any difference. But I wanted to be with her as night fell; I didn't want her to be completely alone. As I sat there the quick tropical dusk gave place to darkness. The stars increased in number and twinkled ever more brightly through the forest canopy. There was a distant pant-hoot far across the valley, but Melissa was silent. Never again would I hear her distinctive hoarse call. Never again would I wander with her from one patch of food to the next, waiting, at one with the life of the forest, as she rested or groomed with one of her offspring. The stars were suddenly blurred and I wept for the passing of an old friend.

The next morning I watched as Melissa took her last, laboured breath: her body shuddered, then relaxed. All around, during those last hours, the branches had swayed and rustled as youngsters played while elders fed on the luscious fruits. *In the midst of life there is death.* This was an appropriate setting for Melissa's passing, allegorical in its portrayal of the inevitable cycles of nature. I was deeply moved, but my tears were over. Melissa had indeed known a hard life, with many misfortunes, but she had lived fully and, for much of the time, had clearly enjoyed living. She had attained a high rank. And, most importantly, she had left a solid succession; Gimble, small but very determined, Gremlin, strong and healthy who would have other infants to carry on her mother's genes, and Goblin, top-ranking male of his community.

16

Gigi

◄◄◄◄◄◄◄◄◄◄◄

G IGI, UNLIKE MELISSA, will leave no descendants. Yet it
would be difficult to overstate the extent to which this large
sterile female has influenced the lives of the Kasakela chimpanzees, particularly the males. Since 1965 when she became sexually
mature she has produced a new pink swelling, more or less regularly,
every thirty days or so. Thus for more than twenty years she has been
almost continually available to the Kasakela males for the gratification
of their sexual desires. During that time her over-worked sex skin has
swelled and shrivelled no less than two hundred and fifty times. By
contrast, Fifi swelled only thirty times during the twenty-year period
following *her* first pinkness. As a result of this repeated and unnatural
stretching Gigi's swelling today is huge when compared with those of
other Gombe females.

Right from the start Gigi radiated sex appeal. Time and again she
has been the nucleus of large and excited sexual gatherings, surrounded by most of all of the males of her community. And once the
adult males have gathered together, drawn by the magnetic presence
of a sexually popular female, they are far more likely to move out to
peripheral areas of their territory to patrol the boundaries. Thus Gigi's
magnificent swelling has, time and again, served as a banner to rally
the Kasakela males, encouraging them to perform valiant deeds in the
protection and expansion of their territory.

In one respect Gigi's sexual popularity is hard to understand, since
she often pulls away from her male partners before the completion of
the sexual act. And she has been doing this for twenty-odd years. I

assume that the males find such behaviour irritating, as well as frustrating, but it has never seemed to dampen their ardour. There are
times, too, when Gigi is extremely reluctant to comply with the sexual
demands of a male and on these occasions her suitors are often remarkably patient. I remember once when Figan was trying to mate
with her. Gigi, who was reclining on the ground, her provocative swelling very much in evidence, totally ignored her suitor's vigorous shaking
of branches. After a few moments Figan, his hair (among other things)
fully erect, stood upright and swayed branches wildly above her re-
cumbent form. Gigi, barely glancing at him, rolled over and lay on
her back, staring at the canopy above. Nonplussed, Figan sat down
for a moment, occasionally shaking a small branch in a jerky, irritated
sort of way as, presumably, he wondered what to do next. Gradually
his branching got more violent, his hair (if it were possible) bristled
even more, and there was a wild gleam in his eye that, I thought, boded
ill for Gigi if she continued to ignore him much longer. Apparently
Gigi got the same message, for she suddenly rose, approached Figan
and crouch-presented before him. But no sooner had he begun to
copulate than she pulled away, screaming, and rushed off.

She then lay down again about ten yards from Figan, who stayed
where she had left him. Presently he lay down too, and all was quiet
for an hour. Then he approached Gigi again — and once more she
utterly ignored his courtship. Not until he repeated his wild, branch-
shaking swagger around her did she finally get up and crouch for
him — but yet again she almost immediately pulled away and ran off.
This time, tight-lipped and scowling, Figan followed and his courtship
was a clear-cut threat. She responded quickly, but the outcome was
the same. Except that Figan, thoroughly stimulated, finally completed
the sexual act — into the air.

There can be no other Kasakela female who has been led off on so
many consortships as Gigi. Time and again she has followed different
males, usually reluctantly, to the various peripheral parts of the home
range that they preferred. She has been, over the past twenty years,
on forty-three such excursions that we know of: the figure is probably
higher. In terms of evolutionary biology the males were 'wasting their
time' since no measure of reproductive success for either partner could

possibly result. However, the males did not know this, so they competed for her favours in all good faith. Moreover, there is little doubt in my mind that even if they *had* understood they would still have voted overwhelmingly in favour of the continued presence of Gigi in their midst.

In one other way Gigi has served the males of her community: she has helped the infants and juveniles to learn the ins and outs of the sexual act. Male chimpanzees are sexually very precocious. From the time they can totter, they show great interest in pink swellings, and they 'mate' pink females zealously throughout their childhood. Of course this is just practice — a male is unlikely to be able to father a child until he is between thirteen and fifteen years of age. But it sometimes seems that Gigi prefers the small sexual advances of infant and juvenile suitors to the more vigorous demands of the adult males. Often she crouches accommodatingly as soon as one of these youngsters starts to court her — approaching with his tiny erection and imperiously shaking a little twig. Indeed, she sometimes actively solicits the sexual attentions of youngsters. Once, for example, she suddenly went over to where Prof and Wilkie were playing a boisterous game, seized Prof by his elbow, pulled him from his playmate, and then, still maintaining her grip, crouched before him. Only when he complied with her wishes did she release him.

At other times she ignores these youngsters completely, however much they persist — and in such matters infants can be surprisingly single-minded for periods of half an hour or more. I remember one long journey when Gigi, fully swollen, was followed by three petulant juvenile suitors. Each of them was quietly whimpering to himself as he followed behind that tempting pink bottom. Each of them approached and shook branches every time she stopped. And each of them was utterly ignored by Gigi.

In 1976 Gigi, for some reason, began to cycle less regularly and, at the same time, became for a while much less popular with the adult males. This may have been due to some hormonal upheaval — they responded to her much as though she was a female showing cycles during pregnancy. And then, one day almost two years later, I happened to be with her when she passed a strange blob of bloody, jelly-like tissue. I preserved it (in whisky, which was the only spirit I had

Melissa had great difficulty transporting her twin sons.

During the early months of the twins' life, Goblin spent more time with his mother than did any other adult male. Their relationship was close and supportive. As they hear an alarming sound, each puts an arm around the other.

The more active twin, Gimble, began to climb onto his mother's back when he was nine months old. But little Gyre stayed listlessly in Melissa's lap.

Gremlin was just eleven years old when she gave birth to her first infant, Getty.

After Getty's birth, Gremlin continued to spend most of her time with Melissa. The characteristic tilt to Melissa's head is the aftermath of her bout with polio, which left her neck partially paralysed.

Getty persistently clung to Gremlin's arm during travel.

Getty chews a piece of palm pith as Gremlin grooms her mother.

Getty had a wonderful relationship with his grandmother, Melissa.

Melissa's last infant, little Groucho, was tiny and listless from birth.

Gigi. (B. Gray)

Gigi displayed much more frequently than most females. Here she charges Melissa, brandishing a frond. Other females prepare to climb a tree to avoid the fracas. (D. Bygott)

Gigi laughs as Patti play-bites her ear. Freud is tickling Gigi's foot.

Gigi presents, encouraging an infant to copulate. (P. McGinnis)

Tita riding Aunt Gigi.

Infants learn a great deal about social behaviour by watching adults. Getty is fascinated as Patti grooms his mother.

An open-mouth kiss as Humphrey and Athena greet each other.

at the time) and sent it to a reproductive biologist. He identified it as a uterine cast such as may, occasionally, be shed (painfully) by human females. What this meant, in Gigi's case, is not known, but afterwards she became slightly more popular with the males — provided there was not too much competition from other females.

With the passing of the years Gigi has become increasingly irritable and unpredictable in her sexual interactions with the younger males. She still, for the most part, responds to their courtship overtures, but often she turns and hits or even attacks them once they start to mate her. There was one occasion when she turned on Prof as he copulated with her in a tree and pushed him so hard that he fell to the rocky ground some twenty feet below. After sitting motionless for a few moments, Prof threw a violent tantrum — to which no one, certainly not Gigi, paid the slightest attention. Incidents of this sort have become ever more frequent and it is scarcely surprising that the young males are less eager than before to mate with this irascible female. What makes it so puzzling is that Gigi seems as keen as ever to *initiate* the sexual act. Again and again she will approach a youthful suitor and solicit copulation. If he avoids her, as is often the case, she usually follows him and tries again. Once, for example, Gigi, fully pink, joined infant Beethoven and his sister, Harmony, as they fed in a tree. Gigi immediately climbed towards Beethoven, but he avoided her. After a few moments she approached once more, but he jumped into another tree. She followed him through that tree and into a third. Then she stopped and began to feed and I thought she had given up. Not a bit of it. After ten minutes or so she climbed over to him yet again, and yet again he avoided her. Gigi pursued a short way, then began to feed once more — until the siblings climbed down and started a grooming session. Gigi followed them at once and hurried after Beethoven when he tried to hide on the far side of his sister. When he then climbed rapidly up a tall tree she sat below, occasionally gazing up at him — wistfully one presumes — for the next thirty minutes. The moment he climbed down she once again approached and crouched, offering him her swelling. And this time her persistence was rewarded — one hour and twenty-five minutes after her first solicitation. That was one time when Gigi neither hit nor threatened her partner!

It is not only infants who are sometimes intimidated by Gigi. Even adolescents are often nervous. For Gigi has become a strong and aggressive female, quite capable of putting most adolescent males firmly in their places. Although it is a fact that male chimpanzees attack more often than females do, this does not mean that there is not an aggressive streak in females. Indeed, many adolescent females go through a highly belligerent phase. But that is before they give birth. Once a female is faced with the task of nurturing a small infant, it ill behoves her to go around swaggering and fighting — she would be putting her precious baby at risk. Thus, most females become less overtly aggressive when they reach maturity.

For Gigi, however, the situation was different, for no infant arrived to temper her naturally assertive, dominant personality. In many ways she now behaves like a male. She has a vigorous charging display, and she displays often. She stands up to threats that most females would avoid and so frequently becomes involved in fights. She represents the ultimate challenge to young males who are desperately trying to dominate the community females. She sometimes accompanies the males on their boundary patrols, not only when she is fully pink but even when she is quite flat. And whereas other females (who only go when they are pink) typically trail along in the wake of the males, Gigi often plays an active role in patrolling activities. She has joined the males in the destruction of the nests of strangers, and in the attacks on females from neighbouring communities. She even took part in some of the brutal assaults during the war with the Kahama chimpanzees.

Gigi has an outstanding hunting record. She takes part in more hunts than other females and has a greater success in capturing prey. She is even able to maintain possession of a kill in the face of vigorous efforts by adult males to steal it. There was, for example, the time when she captured a juvenile colobus monkey, and hung on the carcass like grim death despite three violent attacks by Satan and one by Sherry. During these struggles she fell to the ground three times, locked in combat with Satan, yet managed to escape and, still clutching her prey, rushed up another tree. When Sherry then seized her kill in both hands and pulled as hard as he could, she continued to maintain her grip, even when Satan displayed vigorously around and over both of them. Eventually Sherry managed to tear off the rump and hind legs. Then Gigi was finally able to eat in peace because Satan, rather than continue to try for a share of her meat, opted to follow Sherry and take some of his!

I think the males truly respect this tough and dauntless female, who has been such an integral member of their society for so long. And so, despite her idiosyncratic sexual behaviours, Gigi enjoys relaxed relationships with them and is a favourite grooming partner. Like the males she spends much time in noisy excited social gatherings, whereas most females, unless they are pink, prefer a more peaceful existence, choosing to spend days at a time with family members only and joining the larger groups merely for periodic spells of stimulation. Gigi, again like the males, spends a good deal of time in absolute solitude, whereas other females, after they have had their first baby (provided it lives) never experience real solitude again. For the rest of their lives they are always with one or more of their offspring. Having been a mother myself, I know full well that even a small baby can provide a sense of real companionship.

And so Gigi, in many ways, stands alone. For, despite her many male-like characteristics she is not a male: she has never been, and she never will be, fully integrated into the camaraderie of male society. Nor can she find companionship and comfort, as other females do, within a family. Of course she was part of a family once, but that was long ago. Even when I first knew her, when she was about eight years old, her only relative appeared to be the young male Willy Wally. And he moved away to the south with the Kahama males when the community divided.

With no infant of her own, no opportunity to create, for herself, that special group of close friends, a family unit, Gigi has instead cultivated a number of special relationships with a succession of infants. She became attracted to each of them when the infant was about one and a half years old — the age at which their mothers gave them relative freedom to interact with individuals outside the family circle. When she was with the family, and when the mother permitted it, Gigi would groom, play with or carry her current favourite. She helped to protect the infants too — she was particularly zealous in breaking up play sessions with older youngsters when they began to get too rough. In effect she assumed the role, for one infant after another, of the traditional maiden aunt.

Those were relatively transient relationships, for as the youngsters, at around two and a half years, became increasingly boisterous and self-willed, Gigi lost interest. But more recently she developed some relationships of a more enduring nature — not only with two infants, brother and sister, but also with their mother, Patti. Gigi and Patti spent a fair amount of time together even before Patti gave birth, and afterwards, because of certain inadequacies in Patti's maternal skills, Gigi, for the first time ever, was able to make a really significant contribution to the raising of a child.

Patti immigrated into the Kasakela community in the early seventies, so we knew nothing of her early life. In 1977 her first pregnancy ended in mystery: either the baby was stillborn, or it died during the first few days of its life. At that time Passion and Pom were still hunting newborn infants and Patti's may well have been one of their victims. About a year later she gave birth to an apparently healthy male infant who died as a result of maternal incompetence, for Patti had absolutely no idea how to look after a baby. She did support him with one hand during travel, but sometimes it was his rump that she pressed to her belly so that his head bump, bump, bumped along the ground. Once she trailed him after her by one leg. Sometimes, as she sat feeding, she reached forward for a fruit in such a way that he was crushed between her thigh and belly and made strange high-pitched noises of distress. It was hardly surprising that the baby was dead before the week was out.

A year later Patti gave birth again, to another male whom we named Tapit. Although she was a better mother this time around (which was hardly difficult!) I believe that her infant owed his survival as much to his own tenacity and toughness of spirit as to Patti's care. There were so many times when it seemed that she simply didn't know how to deal with him. Often, for example, she failed to cradle him properly, and then, as she sat grooming or feeding, he would fall back onto the ground. She would let him lie there until he whimpered, whereupon she would quickly gather him up. Once she leapt from one tree to another with Tapit held back to front, his head facing her rear. He screamed loudly during this performance, and when she reached her destination she seemed concerned and sat cradling him — but he was still upside-down, with his feet under her chin and his head in her groin. During his early months incidents of this sort were quite commonplace and Tapit was often to be heard screaming as his mother travelled through the trees.

Because he was cradled so inappropriately, Tapit often had difficulty in reaching Patti's nipples. And in this, most basic of his needs, Patti seemed quite unable to help. As he nuzzled frantically in the wrong place he would whimper, then scream, and though she often looked down at him and watched him intently she almost never adjusted his position to make things easy for him. Even when he finally located a nipple and began to nurse, it was ten to one that a sudden movement of hers would jerk the hard-won reward from his mouth.

By the time he was six months old he was easily able to locate his mother's breasts. But now he was confronted with a new difficulty. One day I followed them to a shady spot in the forest. Patti stretched out to rest and soon Tapit began to suckle. For a few moments all was well — and then Patti began to laugh. I watched in amazement as, chuckling ever louder, she pulled him from her nipple and tickled him, making gentle nibbling movements on his head and face. But Tapit wanted milk, not play. Eventually, whimpering, he succeeded in fighting his way back to the nipple — only to be pulled away again by his still-laughing mother. For some further minutes he tried to get his own way but then gave up, at least for a while. When he suckled next, about an hour later, Patti didn't even try to interrupt, but she showed the same strange play response to suckling on a number of subsequent occasions. Once he struggled for over seven minutes, whimpering all the time, while his mother tickled him and chuckled. Why she behaved in this extraordinary fashion is hard to understand. It is a ploy used by a few mothers during the peak of weaning — they play vigorously with their infants to distract them when they want to suckle or ride during travel. But that is not until the youngsters are about four years old. Patti was obviously confused. Or perhaps it was just that his lips on her nipples tickled, and triggered a play response.

Patti allowed Tapit to move away from her when he was a mere four months old — as soon as he could totter. From this time on, she often left him to his own devices while she groomed or fed nearby.

Sometimes, as he tried to climb towards her up a steep slope or follow her from one branch to another, he would begin to whimper, but she would usually ignore him completely. Often she did no more than glance in his direction, even if he fell a short distance and screamed. She showed the same indifference to his social development. Most mothers are careful to prevent their infants, during the first few months, from making contact with other adults. Not so Patti. When Tapit was only five months old, he climbed onto Satan during a grooming session. Tapit seemed confused and whimpered, but Patti paid no attention. Still whimpering, Tapit clambered across Satan, and soon began to scream. Only then did Patti retrieve him. Another time he tottered away from Patti and climbed a short way up a small sapling. Then he moved over to Gremlin, whimpering. Quickly she embraced him, but he pulled away and stumbled, crying louder, to Gigi. But she had not yet forged a bond with Tapit, and she ignored him. Finally, as he cried louder and louder, Patti, with a slight whimper herself, went over and gathered him up.

When Tapit was nine months old he was subjected to another of his mother's peculiar idiosyncrasies. Again, I was astounded when I saw it for the first time. He was playing in the low branches of a tree near Patti while she fished for termites. When she was ready to leave she stood upright and, instead of putting her hand around his body and gathering him into her embrace in the normal manner, she merely seized one of his ankles and pulled. This, of course, made things very difficult for Tapit. As she continued to tug he clung ever more tightly to his branch, and soon began to scream. Her only response was to pull harder until he was forced to let go — upon which she clamped him to her belly upside-down. This happened again and again during the next couple of months.

By the time Tapit was one year old Patti occasionally wandered off, leaving her son behind. Once, for example, she gradually meandered further and further away from him as she foraged on the sweet yellow fruits of the *budyankende* bushes that cover great tracts of the lower mountain slopes in early summer. She paid no attention to his soft whimpers as he struggled to follow. After a while she was almost out of sight and only when he screamed really loudly did she glance round, then return to gather him up. Four months later she left him on the ground, where he was playing quietly by himself, and climbed a tree to feed. After five minutes Tapit tried to follow his mother, but the climb was too difficult and he began to whimper. Patti did not respond. Even when his cries got louder his mother only looked down at him. Finally Tapit threw a full-blown tantrum, screaming at the top of his voice, hurling himself about on the ground and tearing at his hair. Only then did Patti, somewhat reluctantly, pause in her feeding long enough to retrieve him.

This most unmaternal behaviour meant that, as time went on, mother and child occasionally became separated. Once I met Patti travelling with a group of males: there was no sign of Tapit. When they stopped to feed, Patti fed with them, quite calmly. It was only after fifty minutes that she suddenly seemed to 'remember' that there should be an infant with her! She stopped feeding, looked all around, began to whimper, then ran back the way she had come, crying loudly. I couldn't keep up with her — but later in the day she was seen again, safely reunited with Tapit. Another time when I was following Melissa and her family, we heard the frantic crying of a lost child. At once Gremlin hurried off in the direction of the sounds, and found and embraced the infant — Tapit, of course. She stayed with him, sometimes carrying him, until he found his mother.

When Tapit was just over a year old Gigi began to make overtures of friendship to him. I well remember the first time I saw this. Tapit, as usual, was stumbling along some ten yards behind his mother. It was late in the day when most infants get tired, and even those much older than Tapit typically insist on riding. Presently Tapit began to whimper. Patti, as usual, ignored her son, but Gigi, who had been with them all afternoon, at once went back to him, crouched, and reached her hand towards him, inviting him to climb on her back. He backed away, confused, and lay on his back, crying louder. Gigi moved away at first, but when Tapit got up, still whimpering, she again crouched beside him. And this time Tapit leapt onto her back and she carried him to Patti.

That was the start of the close relationship between them that was to play such a crucial role in Tapit's early development. Gigi, whenever she was not pink, now began to travel with Patti very often, and she lavished any amount of frustrated maternal affection on Tapit. She carried him during travel, groomed him and played with him; she was very protective of him, too. Once an adolescent male baboon, at whom Tapit had been displaying with the bristling, bouncing, exuberance of infancy, suddenly lost patience, grabbed him and rolled him on the ground, then dragged him a short distance. Tapit, only just over a year old, was not surprisingly terrified and began to scream. Patti glanced over, but it was Gigi who sprang into action, raced over and gathered Tapit to her breast. Brave in the presence of his protector, Tapit pulled away from her and again bounced and stamped towards the baboon, hair bristling, while Gigi watched benignly. Then there was the time when Gigi seized Tapit and rushed up a tree just in time to avoid a charge by Goblin. And once, when Satan attacked Patti, causing Tapit, who had been riding on her back, to scream, Gigi actually displayed and kicked out at Satan.

In fact Gigi behaved very much like an elder sister and she and Tapit could often be seen together, sometimes as far as thirty yards from where his mother fed or rested. Once I sat with them during the midday heat while Tapit slept in Gigi's lap for over half an hour, while his mother fed in a tree some distance away. Patti, for her part, seemed delighted with this babysitting arrangement. She showed even less concern for her son when Gigi was around. Once, for example, Tapit went off with Gigi for about a hundred yards while Patti remained grooming with some of the adult males. Her son was quite out of sight, yet even when her group startled and rushed up trees in alarm, Patti seemed totally unconcerned as to Tapit's well-being. Some thirty minutes later he appeared, riding on Gigi's broad back.

During Tapit's third year Patti's treatment of him became, in some ways, even more cavalier than before. During travel he was frequently forced to try to navigate some extremely difficult arboreal routes as he struggled to follow his mother. Even when he screamed she seldom returned to help him. There were many times when he could not cross a gap from one tree to another, despite his most desperate efforts. Then, whimpering and crying the while, he had to climb to the ground and scamper, screaming, to the tree where Patti was calmly feeding. Although youngsters of four and even five years old habitually ride on their mothers' backs in order to cross the fast-running streams,

Patti several times left Tapit on the far bank, forcing him to make his own way over the water via overhanging vegetation, crying loudly. But if Gigi was there to carry or reassure him, all was well. Indeed, she remained a frequent companion, playmate and protectress throughout the rest of his infancy.

There can be no doubt but that Gigi made a tremendous difference to the quality of Tapit's life, bringing him, as she did, concern and reassurance, care and affection. His upbringing was extraordinary and, by the time he was five years old and weaned he was, as might be expected, a remarkable young chimp. He was amazingly independent and self-willed, and yet liable to be thrown into sudden frenzies of anxiety if things went wrong. And then, just before Patti gave birth to her next infant, Tapit died of some unknown disease. How ironic that, having somehow struggled through his perilous infancy, having survived despite his mother, he should leave the world when he was on the very verge of independence.

But Tapit's life was not in vain, for he taught Patti a great deal about maternal behaviour. To my delight she was a wonderful mother to her next infant, daughter Tita, and showed none of the curiously inappropriate behaviour that had characterized her early interactions with Tapit. And so Tapit's tenacity for life will benefit the younger siblings he never met and strengthen Patti's line in future generations of Gombe's chimpanzees.

Gigi began auntying Tita long before she was one year old, presumably because, by then, Patti accepted the big female almost as part of the family. And because of this early start, the relationship between Gigi and Tita was, in some ways, even closer than that between Gigi and Tapit. The bond between the two adult females was growing gradually stronger, too. Indeed, Gigi sometimes became quite upset if she accidentally lost contact with Patti during travelling or feeding.

One day, for example, Gigi climbed to feed about fifteen yards away from Patti and Tita. Some forty minutes later she climbed down and wandered over to the tree where Patti and Tita should have been. But they were not there — they had left a few minutes earlier, moving silently away through the undergrowth. Gigi stared, looked all round, then began to cry and whimper like a child who had lost her mother.

After a few moments she uttered a series of pant-hoots ending in a very loud shout that, at least to my ears, had an exasperated overtone: 'Where on earth have you got to?' A few moments later Patti and her daughter appeared, and the two females groomed for a while. Then Gigi reached towards Tita, gestured to the infant to climb aboard, and set off. Patti had no option but to follow!

I remember vividly another day I spent with the three. After the heat of midday Patti climbed to feed, but Gigi stretched out on the ground and Tita stayed with Gigi. She bounced about, on and around the big female, then began to hit her with a leafy twig. With a play-face, Gigi took one end and they had a tug of war. Then Gigi began to tickle Tita, who responded at once, biting into Gigi's very ticklish neck. Soon both of them were laughing loudly. After ten minutes, Tita had had enough and she climbed to play by herself, swinging in the vines. It was very peaceful. There were a few rustles from the tree where Patti fed and the shrilling of the cicada chorus. Gigi closed her eyes and slept. Suddenly the quiet of the late afternoon was broken as a fight broke out among a nearby troop of baboons. Tita, startled, began to scream and, quick as a flash, Gigi leaped to her feet, rushed up the tree, and gathered Tita to her breast. She carried the infant to the ground and began to groom her until Tita, her eyes closing, relaxed utterly. Then when Patti had finished feeding, the three moved on, Tita riding, carefree and confident, on Aunt Gigi's strong back.

17

Love

≪≪≪≪≪≪≪≪≪≪

POOR GIGI. Unable to bear young of her own, she has not been able to find the sort of reassuring relationships that typically exist between chimpanzee mothers and their grown young. Desperately she has sought contact with a series of infants, but one after the other they have grown away from her. Their bond is with their own mothers — and this bond is the strongest and most meaningful of all. Never again will an individual be so nurtured, protected and cared for as during infancy and early childhood. As the youngster matures, the relationship with the mother strengthens into a close, mutually supportive friendship that may last through life. A male, it is true, may forge a similar friendship with his brother, or even with a non-related community male. But a female, once she loses her own mother (either through death or if she, the daughter, transfers to a new community), will not know such a relationship again until her own youngsters are grown.

The stronger the relationship is between two chimpanzees, the greater the distress if it is threatened. Since the mother is, for her infant, his whole world, it is no wonder that some infants become so depressed during the peak of weaning — for the first time, they experience determined maternal rejection. Throughout the early months of weaning an infant can almost always get his (or her) own way through sheer persistence. But as time passes, his mother prevents him from suckling and from riding on her back with increasing frequency and increasing vigour. Her child's soft and pleading whimpers change, more and more often, to screams of frustration and tantrums. The task of the mother becomes ever harder and, in some cases, is clearly stressful for her as well as for her youngster. This is particularly so if she is trying to wean a firstborn child and is lacking in experience, and even more so if her child is a male, since he is likely to be more violent in his resentment than a female. When, screaming hysterically, he rushes from her and prostrates himself, hitting the ground and pulling at his hair, what is she to do? Usually she follows, often with a grin of fear on her face, and takes hold of him: she wants, I suppose, to calm him. But he, angry and resentful of her rejection, tries to pull away. She, however, maintains her grip, even if he hits or bites her, until he quietens. A female infant often seeks to get her way more subtly, working her way ever closer to the nipple as she grooms her mother, then snatching a quick suck.

At the peak of weaning comes an event that is surely perceived by the infant as yet another threat to the mother–child relationship: the mother resumes her pink swellings. Now, during each oestrus period, she will be preoccupied with the courtship and mating demands of the males and all the attendent commotion. The first couple of cycles are usually the worst, for then the situation is new, strange and frightening for the child. A male infant, as we have seen, tends to interfere in any copulation that takes place nearby. Usually he is quite calm, merely running up and pushing at the male. But when the female is his own mother his interference is often frenzied, and he may hit at the male suitor, grinning and squeaking in his distress. A female infant often seems even more upset when her mother is mated, though she typically ignores the sexual act when non-related females are involved.

We still know little about the correlations between the gradual drying up of the mother's milk, the frequency with which the child suckles, and the hormonal changes in the mother that precede and accompany the development of the next infant in her womb. Some infants suckle throughout the mother's pregnancy. Others are weaned before the mother conceives or during the first few months of gestation. Be that as it may, the birth of the next baby signals the beginning of a new era for the previous child, and it is hardly surprising that some youngsters feel threatened. No longer can they claim the mother's full attention, no longer can they ride on her back or creep into the warm sanctuary of her nest at night. Infancy is left behind. However, although the mother can no longer lavish all her affection on the older child, she is still there to provide reassurance and protection. She will still share food in response to begging. She will groom the older child far more than she grooms the younger. The new-fledged juvenile, therefore, even if upset initially, usually recovers quickly and becomes ever more fascinated by the baby.

Two youngsters did not follow the normal route towards independence, Flint and Michaelmas. Both remained unusually dependent emotionally on their mothers even after the birth of their infant siblings — but for very different reasons. In Flint's case Flo's extreme age seems to have been responsible, for she, who had been the best of mothers in her time, failed this, her lastborn son. If she had not conceived again all would, I think, have gone well for Flint. But that last pregnancy drained so much strength and energy from Flo's aging body that she was simply not able to wean Flint. Surrounded by the assertive members of his high-ranking family, he had become a spoilt and obstreperous child, and when Flo tried to prevent him suckling or riding on her back he threw unusually violent and agressive tantrums. Flo gave in to him again and again, and so he was still nursing right up to the time when little Flame was born. From sheer necessity Flo then managed to wean him from the breast, despite his tantrums, but she could not, it seemed, prevent him from pushing into her nest at night or riding on her back. Indeed, sometimes he insisted on clinging to her belly in the infantile ventral position, utterly obscuring his baby sister. At the same time he became increasingly depressed, playing but seldom and spending long hours sitting close to Flo and grooming her. So it was throughout the six months of his little sister's life. But then Flo fell sick with a pneumonia-like disease. She became so weak that she could not even climb to make a nest at night. And when we found her, lying on the ground, Flame had disappeared, never to be seen again. After she recovered, Flo, still physically and psychologically geared for looking after a small infant, no longer even *tried* to prevent Flint from creeping into bed with her or riding on her back. He did stop riding eventually, but not until he was eight years old, when Flo was simply not strong enough to support his weight.

Michaelmas's story was quite different. He was five years old when his mother, Miff, resumed her pink swellings. During these periods she was very popular and constantly surrounded by many community males. In these large groups, with tension running high, there was, as usual, much aggression, and Miff herself was sometimes attacked. Michaelmas, sticking close to his mother through thick and thin, not only threw himself between his mother and her suitors, but also did his best to interfere when she was attacked. During one such fracas his hip was dislocated. After this, hurting and lame, he could no longer keep up when the family travelled, and Miff, who had been weaning him vigorously before the accident, relented and allowed the child to continue riding on her back. Even after the baby's arrival she continued to carry him often, and when she sometimes ignored his sad whimpering, his elder sister Moeza let him ride on her. Presumably because of his poor physical condition, Miff made no attempt to keep him out of the communal nest, and he continued to join mother and baby. Not until he was seven years old was he observed to make his own night nest and even after that he still occasionally crept into bed with his mother and little sister.

As a youngster gradually becomes more independent, his (or her) relationship with his mother changes. It is still close, the mother is still affectionate and supportive, but the onus of maintaining proximity increasingly falls on the child. Whereas the mother, even if she is ready to move on, will wait for an infant — or go and fetch him if she is impatient — an older child must keep his eye on his mother. This does not mean that she always sets off without her child — far from it. But it does mean that the two will, from time to time, become accidentally separated. When this happens, the child usually becomes very upset. The loud, frenzied screams, interspersed with whimpering calls, that are uttered by lost youngsters are very characteristic. Mothers usually stop and wait when they hear such crying, but for some reason almost never give an answering call. And so the youngster learns two things: first, that he must keep alert to prevent the recurrence of similar experiences; second, that temporary separation from the mother is not, after all, the end of the world — sooner or later they will find one another again. Thus the time eventually comes, earlier for a male than a female, when the child begins deliberately to leave the mother for short periods.

Even after this the youngster is likely to become quite distressed when *accidentally* separated from his mother. Moreover, on those occasions when he and his mother want to travel in different directions, he may try very hard to persuade her to change her mind: if she follows him, separation will be at least temporarily averted. One day in 1982 I was with Fifi and her family: Freud, Frodo and one-year-old Fanni. They had been resting for an hour or so when Freud, eleven years old at that time, sat up, glanced at Fifi, then gathered Fanni to his breast and set off to the north. Fifi, who was grooming Frodo, looked after them, got up and followed. Soon Fanni wriggled free and started back towards her mother, who sat down again, presently rejoined by Freud. After five minutes or so Fifi got up and moved southwards, very slowly, allowing Fanni to totter behind her. Instantly Freud, seizing his chance, reached out to his little sister, gathered her up, and set off once more in the opposite direction. Fifi stopped, gazed after them again, then turned and followed. It was not long before Fanni left Freud, but as she took a few steps towards Fifi, Freud pulled her back and, with little shoves, persuaded her to move on ahead of him. They travelled thus for a few yards, then, as Fanni again tried to escape, Freud grabbed her ankle, drew her close, and groomed her until she relaxed. Fifi just watched. After a couple of minutes, Freud got up and took hold of one of Fanni's arms. Quick as lightning Fifi took hold of the other and pulled gently. Freud soon gave in and Fifi, placing Fanni firmly into the ventral position, set off southward. Freud looked after her for a while, gazed, wistfully perhaps, to the north, then turned and trailed after his family. Much later, as the family was feeding, they heard excited calling from chimpanzees to the east. Freud at once began to travel towards the sounds, but Fifi continued to feed. Freud returned, gathered up Fanni, and set off again. Fifi soon followed. After about seventy-five yards Fanni dropped down and went back to her mother but this time Fifi went with Freud, and the family joined the large group.

All of the above — weaning, the birth of a new baby, temporary separation — upsetting as they may be at the time, are as nothing when compared with the death of the mother, the final and irrevocable breaking of the bond. Infants who are less than three years old and still quite dependent on their mother's milk will, of course, be unable to survive. But even youngsters who are nutritionally independent may become so depressed that they pine away and die. Flint, for example, was eight and a half when old Flo died, and should have been able to look after himself. But, dependent as he was on his mother, it seemed that he had no will to survive without her. His whole world had revolved around Flo, and with her gone life was hollow and meaningless. Never shall I forget watching as, three days after Flo's death, Flint climbed slowly into a tall tree near the stream. He walked along one of the branches, then stopped and stood motionless, staring down at an empty nest. After about two minutes he turned away and, with the movements of an old man, climbed down, walked a few steps, then lay, wide eyes staring ahead. The nest was one which he and Flo had shared a short while before Flo died. What had he thought of as he stood there, staring? Memories of happy days gone by to add to his bewildered sense of loss? We shall never know.

It was unfortunate that, for the first few days after Flo's death, Fifi had been wandering further afield. Had she been there to comfort Flint from the start, things might have been very different. He had travelled for a while with Figan and, in the presence of his big brother, had seemed to shake off a little of his depression. But then he suddenly left the group and raced back to the place where Flo had died and there sank into ever deeper depression. By the time Fifi showed up Flint was already sick, and though she groomed him and waited for him to travel with her, he lacked both the strength and the will to follow.

Flint became increasingly lethargic, refused most food and, with his immune system thus weakened, fell sick. The last time I saw him alive, he was hollow-eyed, gaunt and utterly depressed, huddled in the vegetation close to where Flo had died. Of course, we tried to help him. I had to leave Gombe soon after Flo's death, but one or other of the students or field assistants stayed with Flint each day, keeping him company, tempting him with all kinds of foods. But nothing made up for the loss of Flo. The last short journey he made, pausing to rest every few feet, was to the very place where Flo's body had lain. There he stayed for several hours, sometimes staring and staring into the water. He struggled on a little further, then curled up — and never moved again.

Other youngsters, though, have been cared for by their older siblings. And these adoptions provide us with some of the most touching stories, illustrating clearly the nature of the affectionate, protective attitude of juveniles and adolescents towards their infant brothers and sisters. Young males, it transpires, can be just as efficient caretakers as females. Certainly they are as tolerant and affectionate. This first became clear in Passion's family.

Pax was but four years old when his mother died. She had been ill for some weeks, moving more and more slowly, becoming increasingly emaciated, crouching to the ground from time to time as though in pain. Though I had hated her four years before, during her infant-killing days, I could not help but feel sorry for her at the ending of her life. On the last evening she was so weak that she trembled when she made the slightest movement. She managed to climb into a low tree where she made a tiny nest, then lay, exhausted. The next morning dawned cold and grey with rain pouring steadily from a leaden sky. Passion was dead. She had fallen during the night and hung, caught by one arm, from a tangle of vines. Her three offspring, who had been her constant companions during the last weeks of her life, were around her now. Pom and Prof, for the most part, just sat staring at their mother's body. But Pax repeatedly approached and tried to suckle from her cold, wet breasts. Then, becoming increasingly upset, screaming louder and louder, he began to pull and tug at her dangling hand. So frenzied was he in his distress that eventually he succeeded in pulling her loose. As Passion sprawled lifeless on the sodden ground, her three offspring inspected her body many times. Occasionally they moved off a short distance to feed, listlessly, then hurried back to their dead mother again. As the day wore on Pax gradually became calmer and no longer tried to suckle, but he seemed ever more depressed, crying softly and, occasionally, pulling at Passion's dead hand. Eventually, just before darkness fell, the three moved off together.

For the next few weeks Pax showed many signs of depression. He was listless, he played not at all, and, like all young orphans, he soon developed a pot belly. But he recovered amazingly fast. For about a year the three siblings spent almost all their time together. When Prof ventured to travel for a while with the adult males, Pax usually stayed with Pom. But although they kept close together, and although he invariably ran to her for protection, Pax for some strange reason would never ride on his sister: not even when, as they travelled with a group of fast-moving adult males, Pax got left behind and whimpered; not even when she reached out, begging him to climb aboard. Initially, her maternal instincts roused, Pom had tried to *force* him to climb her back. But Pax had clung to the vegetation, and screamed hysterically until she stopped. Prof had tried to carry his little brother too, but Pax had rejected those offers in the same inexplicable way. And it was just the same when his elder siblings invited him to share their nests at night. He utterly refused, even when they reached affectionately towards him. And so they had watched as Pax, whimpering sadly to himself, made his own small nest nearby. How much we have yet to learn.

One year after Passion died, Pom emigrated and joined the Mitumba community in the north. She probably did so because, after losing her high-ranking mother, she had been at the mercy of the Kasakela females many of whom, without doubt, retained hostile feelings towards her: chimpanzees have long memories. But even before his sister left, Pax had attached himself to his brother, following Prof, like a small persistent shadow, wherever he went. The relationship between the two had always been affectionate, for Prof had been fascinated by Pax from the start, and had often carried and played with his little brother. I remember once when Pax, suffering from a wet-season cold, sneezed loudly and messily. Prof hastened over and gazed intently at Pax's runny nose — then picked a handful of leaves and carefully wiped the snot away.

Now, a year after Passion's death, Prof, in many ways, cared for Pax as a mother would, waiting for him in travel and protecting him. Even when Pax was six years old he became extremely upset if he was accidentally separated from Prof. And Prof was concerned, too. Once, for example, a full two years after losing their mother, the brothers went off in different directions when the big group in which they had been feeding split up. When Pax noticed that Prof was not there, he began to whimper and cry. Repeatedly he climbed tall trees, crying louder and scanning the countryside all round. But Prof by then was out of sight and hearing, and so Pax stayed close to Jomeo, making his nest close to that of the big male. Even so he cried, on and off, throughout the night. Prof, for his part, left the other chimps as soon as he realized what had happened and set off to search for Pax. I didn't see the reunion, but by noon the next day they were together again.

One incident I shall always remember. The brothers were travelling in a small group with Miff, who was pink, and Goblin, who was jealously asserting his rights as alpha and preventing other males from mating with her. He paid no attention when Pax courted Miff — the youngster was no threat. Miff, however, seemed irritated by the courtship of this puny suitor and when he persisted she kicked back at him. He was hurled head over heels into the vegetation behind him. Poor Pax! He threw one of the most violent tantrums I have ever seen. Tearing at his hair, he threw himself about on the ground, screaming louder and louder. Goblin, obviously irritated by the noise, glared at Pax, and his hair began to bristle. At that moment Prof, who had been feeding some distance away, came hurrying up to see what was going on. For a moment he stood surveying the scene then, realizing that Pax was in imminent danger of severe punishment, seized his still-screaming kid brother by one wrist and dragged him hastily away! Not until they had gone at least twenty yards and were well out of danger did Prof let go: at that point Pax stopped screaming and agreed to go off with his brother.

Gimble was eight years old when Melissa died and, although still tiny for his age, was well able to fend for himself. Even so, he was upset and a little dazed when he lost his mother. He turned to his siblings for comfort and, of the two, it was Goblin whom he sought most often, and soon he was following his elder brother everywhere. Often they fed side by side in the same tree and Gimble made his night nest close to Goblin's. Most important, from Gimble's point of view, Goblin usually supported his small brother if he were threatened or attacked by any of the others. Thus Goblin, alpha male and thirteen years older than his brother, in many ways filled Melissa's place in Gimble's life.

When Winkle died, Wolfi was adopted by his elder sister, Wunda: the story of the nine-year-old female and her three-year-old brother is truly remarkable. Wolfi, despite his young age, showed fewer signs of depression than the other orphans and almost certainly this was because, long before Winkle died, the relationship between the siblings had been unusually close. Wunda had carried him frequently when the family travelled, not only because, like all elder sisters, she was fascinated with her small brother, but also because, from the time he was able to totter, Wolfi had wanted to follow her wherever she went. Again and again Wunda had set off about her own concerns, only to return when she heard the sad cries of her small brother as he tried, most desperately, to keep up. Then she would gather him up, and off they would go, together. It should not be thought that Wolfi's close relationship with his sister reflected adversely on Winkle's maternal abilities: she was a caring, affectionate and efficient mother from whom Wunda, undoubtedly, had learned much concerning child care. When Winkle died Wunda took over all her caretaking duties as a matter of course. Most amazing of all, this young female, not yet sexually mature, may have actually produced milk for her infant brother. Certainly he suckled, for several minutes every couple of hours or so, and he became very upset if Wunda tried to stop him. But even when we got very close to them we still couldn't be sure that he was actually getting milk from his sister. Perhaps he just found it reassuring to put his lips to her nipples.

Skosha was a firstborn child and had no brother or sister to care for her when her mother died. For the first two months, this five-year-old spent most of her time with one or other of the adult males. But then she became attached to Pallas, a female who had, a few months earlier, lost her own first child. Pallas had been a very close companion of Skosha's mother, and we had often wondered whether, perhaps, the two were sisters — if so, then Pallas was Skosha's biological aunt. Be that as it may, the two became inseparable. Pallas was a wonderful foster mother. She carried Skosha during travel, waited for her, shared food, and was remarkably patient with this child, who, when things went wrong, often threw violent tantrums. Within the year Pallas again gave birth — to an infant who, almost certainly, fell victim to Passion and Pom. The following year, however, Pallas had another baby, who survived, and by that time Skosha was a fully integrated member of the family. And it was a delightful family, too, for Pallas, although she was not a very social female, was an affectionate and playful mother, and little Kristal, outgoing, adventurous and tough, became a favourite of us all. But ill luck dogged Pallas: she fell sick and died when Kristal was just five years old. And so Skosha, having lost her own mother, now lost her foster mother too.

I arrived at Gombe soon afterwards. It was heartbreaking to see the two orphans. Skosha was doing her best to look after Kristal, but the child was depressed and lethargic, and Skosha herself, now ten years old, seemed forlorn and bewildered. She clearly found it difficult to decide on any course of action. Where should they go next? What should they eat? When should they make their nests? Kristal kept very close to Skosha as the two of them wandered aimlessly through the forest, two lost babes in the wood. We all hoped that Kristal would survive, but she remained listless and never recovered her former gay spirit. Nine months after Pallas died, Kristal disappeared for good.

In 1987 an epidemic of a pneumonia-like disease swept through the Gombe chimpanzee population. Many members of the Kasakela community fell sick, and although some, like Evered and Fifi and Gremlin, made wonderful recoveries, nine chimps died. Jomeo, Satan and Little Bee were among my oldest friends to go. Another was Miff, whom I had known since she was a juvenile in 1964. Just a few years before she died, Miff had a flourishing family. But first Michaelmas (whose limp, incidentally, had quite gone), became sick and died of a heavy infestation of internal parasites. And then juvenile Mo had vanished after a long sickness. And now Miff herself was gone, leaving a sickly three-year-old, little Mel. He was all alone in the world — Miff's eldest offspring, daughter Moeza, was still alive, but she had emigrated, three years earlier, to the Mitumba community.

I was in the States on my annual spring lecture tour when I had a letter from Gombe telling me the news. Mel, I heard, was very weak. He was wandering around after various individuals, mostly one or other of the adult males and, although all were tolerant, none showed special concern. I never expected to see Mel again. Even before Miff's death he had been so skinny and pot bellied and lethargic that I had sent a faecal sample to be analysed, and the report, listing very heavy infestations of several different types of internal parasites, had not been encouraging. But then I had a telegram — *Mel adopted by Spindle*. I was amazed, for Spindle, twelve-year-old son of old Sprout, was quite unrelated to Miff so far as we knew. Surely such a relationship couldn't last?

Soon after that I returned to Gombe to find Mel still alive, and still with Spindle. Looking at the little orphan, with his pot belly, his skinny arms and legs, his dull, sparse hair, I marvelled at the gallant fighting spirit that had enabled him, against all odds, to cling to life. I marvelled, too, at the concern and affection shown by his caretaker. Spindle was in a sense an orphan himself, as Sprout had died during the same epidemic that had claimed Miff and so many others. Spindle, of course, was well able to care for himself: but was it, perhaps, his sense of loss, a feeling of loneliness, that had led to his unlikely liaison with an unrelated motherless infant? Whatever the reason, Spindle was a wonderful caretaker. He shared his night nest with Mel, and he shared his food. He did his best to protect the infant, hurrying to retrieve him when the big males became socially aroused. When Mel whimpered during travel, Spindle waited and allowed him to clamber onto his back or even, if it was raining and cold, to cling on in the ventral position. In fact he carried him so often that, where Mel gripped Spindle with his feet, the hair became quite worn away and Spindle had two large, white hairless patches, one on each loin!

The main problem that Mel had to cope with, in addition to the loss of his mother, his heavy parasite load and general malaise, was the fact that Spindle was travelling with the adult males and, at that time of year, they were moving long distances each day, searching for fallen *mbula* fruits. They often went out to the northern periphery of their range during these feeding excursions, and several times, after hearing calls made by males of the powerful Mitumba community, they travelled silently, and very fast, back towards the centre of their home range. It was tough for little Mel, because Spindle, patient as he was, did not always wait for his small charge. Mel had to cover a great deal of ground on his own.

Most of the other chimps, particularly the adult males, were amazingly gentle and tolerant in their interactions with the orphan. He was able to approach any one of them, without fear, to beg for food — even pushing in to take meat after a kill when tension runs high among competing individuals. At most, Mel's presumption elicited a mild threat — which invariably caused *him* to throw a tantrum! And often he was successful in his attempts to get a share.

Towards the end of July, Spindle and Mel became separated. Mel was very distressed. For a few days he followed one or other of the adult males, even, during sudden excitement, jumping onto their backs. And then he found a temporary substitute for Spindle. Incredibly, it was Pax who took him on.

It was five years after Passion's death, and Pax was ten years old, but, like almost all the orphans that survive the loss of their mothers, very tiny for his age. He was still inseparable from Prof, the bond between them as strong as ever. I shall never forget that summer, and the days I spent with the two brothers and little Mel. Prof almost always led travel while Pax, with Mel clinging tightly to his back, stumped behind his brother along the forest trails and over the streams. He even carried his charge part way up some of the bigger trees. It was not long before Pax developed those badges of service — two white hairless patches, one on each loin! Like Spindle, Pax shared his nest and his food with Mel. And Prof sometimes shared *his* food with both of them! Although it seemed that these three had become very close, after a few weeks Mel was reunited with Spindle and the two remained inseparable for several more months.

A year after losing his mother Mel seemed a little healthier: his arms and legs were not quite so stick-like, his belly not quite so rotund, and his hair was thicker and glossier. He was also less depressed, less withdrawn, and he occasionally joined another youngster for a gentle game. Even though his improved health was due in part to the fact that we managed to give him some medication for his parasites, there is little doubt that Mel survived because of the care he received from Spindle. By the time he was four years old, however, Mel began to spend less time with his benefactor, and gradually, during the following year, the bond between them lessened.

This was when Mel began to travel, more and more often, with Gigi. And with them, almost always, was Darbee, whose mother, Little Bee, had died in the same epidemic that had claimed Miff. Darbee has an elder brother, and I had expected him to care for her, but although she had spent much time with him during the weeks immediately following the death of their mother, the two never became really close. Instead, Darbee formed temporary attachments to two adolescents, one male and one female, before taking up with Gigi. As time went on, it became commonplace to see Gigi, Darbee and Mel together, the large, childless female in the lead, the two small motherless infants following behind.

Gigi's relationship with these orphans is of a different nature from those that she forged with younger infants in the past. In those cases it was Gigi who desired the association: she had to work not only to attract the infants themselves, but also to curry favour, to some extent, with their mothers. Now, however, it is Mel and Darbee who have chosen to attach themselves to Gigi. Gigi shows them little overt affection, and their friendly interactions are, for the most part, confined to occasional grooming. But she provides the support they need in an often unfriendly world. Woe betide any boisterous juvenile or adolescent whose rough behaviour causes one of her small wards to scream — they have Gigi to contend with. When the orphans are with her they can, to some extent, relax, knowing that she will make all decisions regarding travel routes, sleeping places, and so on. But when Gigi is pink and travelling with the big males, Mel and Darbee do not always follow her, preferring to stay on their own, away from the excitement and commotion of the larger groups.

These two infants have survived, but the psychological scars of their ordeal will never leave them. You can tell when you look into their eyes — they lack the sparkle and eager curiosity of normal youngsters of their age. In many ways they behave like adults: their movements are deliberate, and they spend much time resting and grooming themselves. They seldom play, and when they do it is not the exuberant and boisterous play normal for their age but quiet and sedate, and they are quick to fancy hurt when none was meant. How will they behave as adults, they and the others who have suffered similar traumas in their early years? We can only find the answers by waiting, patiently waiting and watching and recording. When I first arrived at Gombe, field studies that lasted as long as one year were almost unheard of. Louis Leakey predicted it would take ten years to begin to understand the chimpanzees. How delighted he would be to see the research that grew from his wisdom moving into its fourth decade.

18

Bridging the Gap

‹‹‹‹‹‹‹‹‹‹‹‹

LOUIS LEAKEY sent me to Gombe in the hope that a better understanding of the behaviour of our closest relatives would provide a new window onto our own past. He had amassed a wealth of evidence that enabled him to reconstruct the physical characteristics of early humans in Africa, and he could speculate on the use of the various tools and other artifacts found at their living floors. But behaviour does not fossilize. His curiosity about the great apes was due to his conviction that behaviour common to modern man and modern chimpanzee was probably present in our common ancestor and, therefore, in early man himself. Louis was way ahead of most of his contemporaries in his thinking, and today his approach seems even more worthwhile in view of the surprising discovery that, as mentioned, human DNA differs from chimpanzee DNA by only just over one per cent.

There are many similarities in chimpanzee and human behaviour — the affectionate, supportive and enduring bonds between family members, the long period of childhood dependency, the importance of learning, non-verbal communication patterns, tool-using and tool-making, cooperation in hunting, sophisticated social manipulations, aggressive territoriality, and a variety of helping behaviours, to name but a few. Similarities in the structure of the brain and central nervous system have led to the emergence of similar intellectual abilities, sensibilities and emotions in our two species. That this information concerning the natural history of chimpanzees has been helpful to those studying early man is demonstrated, again and again, by the frequency with which anthropological textbooks refer to the behaviour of the Gombe chimpanzees. Of course, theories regarding the behaviour of early man can never be anything but speculative — we have no time-machine, we cannot project ourselves back to the dawn of our species to watch the behaviour and follow the development of our forebears: if we seek to understand these things a little we must do the best we can with the flimsy evidence available. So far as I am concerned, the concept of early humans poking for insects with twigs and wiping themselves with leaves seems entirely sensible. The thought of those ancestors greeting and reassuring one another with kisses or embraces, cooperating in protecting their territory or in hunting, and sharing food with each other, is appealing. The idea of close affectionate ties within the Stone Age family, of brothers helping one another, of teen-age sons hastening to the protection of their old mothers, and of teen-age daughters minding the babies, for me brings the fossilized relics of their physical selves dramatically to life.

But the study at Gombe has done far more than provide material upon which to base our speculations of prehistoric human life. The opening of this window onto the way of life of our closest living relatives gives us a better understanding not only of the chimpanzee's place in nature, but also of *man's* place in nature. Knowing that chimpanzees possess cognitive abilities once thought unique to humans, knowing that they (along with other 'dumb' animals) can reason, feel emotions and pain and fear, we are humbled. We are not, as once we believed, separated from the rest of the animal kingdom by an un-bridgeable chasm. Nevertheless, we must not forget, not for an instant, that even if we do not differ from the apes in kind, but only in degree, that degree is still overwhelmingly large. An understanding of chimpanzee behaviour helps to highlight certain aspects of human behaviour that *are* unique and that *do* differentiate us from the other living primates. Above all, we have developed intellectual abilities which dwarf those of even the most gifted chimpanzees. It was because the gap between the human brain and that of our closest living relative, the chimpanzee, was so extraordinarily large, that palaeontologists for years hunted for a half-ape, half-human skeleton that would bridge this human/non-human gap. In fact, this 'missing link' is comprised of a series of vanished brains, each more complex than the one before: brains that are for ever lost to science save for a few faint imprints on fossil craniums; brains that held, in their increasingly intricate convolutions, the dramatic serial story of developing intellect that has led to modern man.

Of all the characteristics that differentiate humans from their non-human cousins, the ability to communicate through the use of a sophisticated spoken language is, I believe, the most significant. Once our ancestors had acquired that powerful tool they could discuss events that had happened in the past and make complex contingency plans for both the near and the distant future. They could teach their children by explaining things without the need to demonstrate. Words gave substance to thoughts and ideas that, unexpressed, might have remained, for ever, vague and without practical value. The interaction of mind with mind broadened ideas and sharpened concepts. Sometimes, when watching the chimpanzees, I have felt that, because they have no human-like language, they are trapped within themselves. Their calls, postures and gestures, together, add up to a rich repertoire, a complex and sophisticated method of communication. But it is non-verbal. How much more they might accomplish if only they could *talk* to each other. It is true that they can be *taught* to use the signs or symbols of a human-type language. And they have cognitive skills to combine these signs into meaningful sentences. Mentally, at least, it would seem that chimpanzees stand at the threshold of language acquisition. But those forces that were at work when humans began to speak have obviously played no role in shaping chimpanzee intellect in this direction.

Chimpanzees also stand at the threshold of another uniquely human behaviour — war. Human warfare, defined as *organized armed conflict between groups*, has, over the ages, had a profound influence on our history. Wherever there are humans they have, at one time or another, waged some sort of war. Thus it seems most likely that primitive forms of warfare were present in our earliest ancestors, and that conflict of this sort played a role in human evolution. War, it has been suggested, may have put considerable selective pressure on the development of intelligence and of increasingly sophisticated cooperation.

The process would have escalated — for the greater the intelligence, cooperation and courage of one group, the greater the challenge to its enemies. Darwin was among the first to suggest that warfare may have exerted a powerful influence on the development of the human brain. Others have postulated that warfare may have been responsible for the huge gap between the human brain and that of our closest living relatives, the great apes: hominid groups with inferior brains could not win wars and were exterminated.

Thus it is fascinating as well as shocking to learn that chimpanzees show hostile, aggressive territorial behaviour that is not unlike certain forms of primitive human warfare. Some tribes, for example, carry out raids during which 'they stalk or creep up to the enemy, using tactics reminiscent of hunting' — thus writes Renke Eibl-Eibesfeldt, an ethologist who has studied aggression in peoples around the world. Long before sophisticated warfare evolved in our own species, pre-human ancestors must have shown preadaptations similar — or identical — to those shown by the chimpanzees today, such as cooperative group living, cooperative territoriality, cooperative hunting skills, and weapon use. Another necessary preadaptation would have been an inherent fear or hatred of strangers, sometimes expressed by aggressive attacks. But attacking adult individuals of the same species is always a dangerous business and, in human societies, in historical times, it has been necessary to train warriors by cultural means such as glorifying their role, condemning cowardice, offering high rewards for bravery and skill on the battle field, and emphasizing the worthiness of practising 'manly' sports during childhood. Chimpanzees, however, particularly young adult males, clearly find inter-group conflict attractive, despite the danger. If young male prehumans also found excitement in encounters of this sort, this would have provided a firm biological basis for the glorification of warriors and warfare.

Among humans, members of one group may see themselves as quite distinct from members of another, and may then treat group and non-group individuals differently. Indeed, non-group members may even be 'dehumanized' and regarded almost as creatures of a different species. Once this happens people are freed from the inhibitions and social sanctions that operate within their own group, and can behave to non-group members in ways that would not be tolerated amongst their own. This leads, among other things, to the atrocities of war. Chimpanzees also show differential behaviour towards group and non-group members. Their sense of group identity is strong and they clearly know who 'belongs' and who does not: non-community members may be attacked so fiercely that they die from their wounds. And this is not simple 'fear of strangers' — members of the Kahama community were familiar to the Kasakela aggressors, yet they were attacked brutally. By separating themselves, it was as though they forfeited their 'right' to be treated as group members. Moreover, some patterns of attack directed against non-group individuals have never been seen during fights between members of the same community — the twisting of limbs, the tearing off of strips of skin, the drinking of blood. The victims have thus been, to all intents and purposes, 'dechimpized', since these are patterns usually seen when a chimpanzee is trying to kill an adult prey animal — an animal of another species.

Chimpanzees, as a result of an unusually hostile and violently aggressive attitude towards non-group individuals, have clearly reached a stage where they stand at the very threshold of human achievement in destruction, cruelty and planned inter-group conflict. If ever they develop the power of language, might they not push open the door and wage war with the best of us?

What of the other side of the coin? Where do the chimpanzees stand, relative to us, in their expression of love, compassion and altruism? Because violent and brutal behaviour is vivid and attention-catching, it is easy to get the impression that chimpanzees are more aggressive than they really are. In fact, peaceful interactions are far more frequent than aggressive ones; mild threats are more common than vigorous ones; threats per se are much more frequent than fights; and serious, wounding conflicts are rare compared to brief, relatively mild ones. Moreover, chimpanzees have a rich repertoire of behaviours that serve to maintain or restore social harmony and promote cohesion among community members. The embracing, kissing, patting and holding of hands that serve as greetings after separation, or are used by dominant individuals to reassure their subordinates after aggression. The long, peaceful sessions of relaxed social grooming. The sharing of food. The concern for the sick or wounded. The readiness to help companions in distress, even when this means risking life or limb. All these reconciliatory, friendly, and helping behaviours are, without doubt, very close to our own qualities of compassion, love and self-sacrifice.

At Gombe care of the sick is not a helping behaviour common among unrelated chimpanzees. Indeed, a badly injured individual is sometimes shunned by non-family members. When Fifi, who had a gaping wound in her head, repeatedly solicited grooming from others in her group, they peered at the injury (where some fly maggots could be seen) then moved hastily away. But her infant son groomed carefully around the edges of the lesion and sometimes licked it. And when old Madam Bee lay dying, after the assault by the Kasakela males, Honey Bee spent hours each day grooming her mother and keeping the flies away from her terrible wounds. In groups of captive chimpanzees, individuals who have been raised together, and who are as familiar as close kin in the wild, will zealously squeeze or poke pus from one another's wounds and remove splinters. One took a speck of grit from his companion's eye. A young female developed the habit of cleaning her companion's teeth with twigs. She found this particularly fascinating when their milk teeth were loose and wiggly, and she even performed a couple of extractions! Such manipulations are for the most part due to a fascination for the activity itself, and almost certainly derive from social grooming. The results, however, are sometimes beneficial to the recipients and, coupled with the concern so often shown for family members, the behaviour provides a biological base for the emergence of compassionate health care in man.

Among non-human primates in the wild it is rare for adults to share food with each other, although mothers will typically share with their young. In chimpanzee society, however, even non-related adults frequently share with each other, although they are more likely to do so with kin and close friends. At Gombe sharing among adults is seen most often during meat eating when, in response to an outstretched hand or other begging gesture, the possessor may allow a portion of the flesh to be taken — or may actually tear off a piece and hand it to the supplicant. Some individuals are much more generous than others in this respect. Sometimes other foods in short supply are shared, too — such as bananas. A good deal of sharing is seen among captive chimps. Wolfgang Kohler, 'in the interests of science', once shut the young male Sultan into his cage without his supper, while feeding the old female Tschego outside. As she sat eating her meal, Sultan became increasingly frenzied in his appeals to her, whimpering, screaming, stretching his arms towards her, and even throwing bits of straw in her direction. Eventually (when, presumably, she had taken the edge off her own hunger) she gathered a pile of food together and pushed it into his cage.

Food sharing among chimpanzees is usually explained away by scientists as being merely the best way of getting rid of an irritation — the begging of a companion. Sometimes this is undoubtedly true, for begging individuals can be extraordinarily persistent. Yet often the patience and tolerance of the individual who has possession of the desired object is remarkable. There was, for example, the occasion when old Flo wanted the piece of meat that Mike was chewing. She begged with both hands cupped around his muzzle, for well over a minute. Gradually she moved her pouted lips closer and closer until they were within an inch of Mike's. In the end he rewarded her, pushing the morsel (well chewed by then) directly from his mouth into hers. And what of Tschego's feeding of young Sultan? Admittedly she may have been irritated by his noisy tantrum — but she could have walked away to the far corner of her enclosure. Robert Yerkes tells of a female who was offered fruit juice from a cup through the bars of her cage. She filled her mouth and then, in response to pleading whimpers from the next cage, walked over and transferred the juice into her friend's mouth. She then returned for another mouthful which she delivered in the same manner. And so it continued until the cup was empty.

Towards the end of Madam Bee's life there was an unusually dry summer at Gombe, and the chimpanzees had to travel long distances between one food source and the next. Madam Bee, old and sick, sometimes got so tired during these journeys that she had no energy to climb for food upon arrival. Her two daughters would utter soft calls of delight and rush up to feed, but she simply lay below, exhausted. On three quite separate occasions Little Bee, the elder daughter, after feeding for about ten minutes, climbed down with food in

Flint insisted on riding Flo even after Flame was born. (Hugo van Lawick)

Flint threw violent tantrums when Flo prevented him from suckling. (Hugo van Lawick)

Flint seemed unable to survive after Flo's death, even though he was over eight years old at the time. He showed signs of depression and, in a weakened state, sickened and died three weeks after his mother. (Hugo van Lawick)

Passion was very ill and clearly in pain for several weeks prior to her death. Four-year-old Pax tries to suckle as Prof grooms his mother.

After her adoptive mother's death, Skosha tried to care for her five-year-old foster sister, Kristal.

There had been a very close bond between Wunda and her three-year-old brother Wolfi even before their mother died. Perhaps as a result of this, Wolfi showed fewer signs of depression than other young orphans. Wunda acts in a very maternal way.

Little Mel was just three, and very sickly, when Miff left him an orphan.

Three weeks after losing his mother, Mel had bonded with a twelve-year-old adolescent male, Spindle, who cared for him as a mother will for her infant. Note the hairless patch on Spindle's groin where Mel so often gripped with his feet. (T. Collins)

Gigi with the two orphans she has adopted, Mel and Darbee.

Fifi with her offspring: Fanni, Frodo, and Freud. A close and enduring family bond.

Charlie, rescued by Simon and Peggy Templar, tries to give himself a 'fix' with the syringe used to sedate him on the journey to sanctuary in Britain. *(The Star)*

Chimpanzees at SEMA Inc., Rockville, Maryland. This photo was taken in 1988. Note the door of the 'isolette' which, when closed, deprives youngsters of contact with the outside world. (© People for the Ethical Treatment of Animals)

Desperate for contact with another living being, JoJo tries to groom my hand. This fully adult male has been imprisoned in his five-by-five-foot, seven-foot-high cell for over ten years. (Susan Farley)

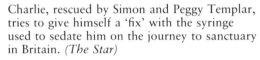

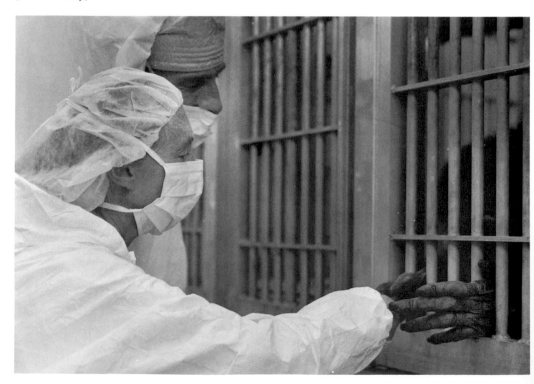

Steve Matthews with one of the chimps used by photographers on the beaches in the south of Spain. Matthews was trying to discover how many chimps were involved. This little chimp has a mark below his left eye, where he has been repeatedly burned with a cigarette.

Whiskey has been kept on a two-foot chain at the back of a garage in Burundi since the age of about six, when he became too strong to continue living in his owner's house. He was bought from the hunters who shot his mother for food. (Steve Matthews)

Whiskey and I share a hug.
(Steve Matthews)

Mike. (Hugo van Lawick)

Fifi fondling
little Faustino.

Evered in the forest.

her mouth and food in one hand, then went and placed the food from her hand on the ground beside Madam Bee. The two sat side by side, eating companionably together. Little Bee's behaviour was not only a demonstration of entirely voluntary giving, but it also showed that she understood the needs of her old mother. Without understanding of this sort there can be no empathy, no compassion. And, in both chimpanzees and humans, these are the qualities that lead to altruistic behaviour and self-sacrifice.

In chimpanzee society, although most risk-taking is on behalf of family members, there are examples of individuals risking injury if not their lives to help non-related companions. Evered once risked the fury of adult male baboons to rescue adolescent Mustard, pinned down and screaming, during a baboon hunt. And when Freud was seized during a bushpig hunt by an enraged sow, Gigi risked her life to save him. The pig had seized him from behind, and Freud, dropping his piglet, was screaming and struggling to escape, when Gigi raced up, hair bristling. The sow wheeled to charge Gigi, and Freud, bleeding heavily, was able to escape up a tree.

In some zoos, chimpanzees are kept on man-made islands, surrounded by water-filled moats. There are tales of heroism from there, too. Chimpanzees cannot swim and, unless they are rescued, will drown if they fall into deep water. Despite this, individuals have sometimes made heroic efforts to save companions from drowning — and were sometimes successful. One adult male lost his life as he tried to rescue a small infant whose incompetent mother had allowed it to fall into the water.

In all those animal species in which parents devote time and energy to the raising of their young they will, when occasion demands, risk life and limb in defence of their offspring. It is much more unusual for an adult to show altruistic behaviour towards an individual who is not closely related. After all, if you help your kin, all of whom bear some of the same genes as yourself, then your action will still be of some benefit to your clan in its struggle to survive — even if you yourself get harmed in the process. From these basically selfish roots sprang the most rarified form of altruism — helping another even when you stand to gain nothing for yourself or your kin.

As the ancestors of chimpanzees (and, incidentally, ourselves) gradually evolved more complex brains, so the period of childhood dependency became longer and mothers were forced to expend more
and more time and energy in raising their families. Mother–offspring
bonds became ever more enduring. The offspring of the most caring,
supportive and successful mothers thrived and became themselves
good and caring mothers who tended to produce many offspring.
Youngsters who were less well cared for were less likely to survive,
and those that did were often relatively poor mothers themselves and
were less likely to produce large families. Loving and nurturing characteristics thus competed successfully, in the genetic sense, with more
selfish behaviours. Over the aeons, tendencies to help and protect,
which were originally developed for the successful raising of young,
gradually infiltrated the genetic make-up of chimpanzees. Today we
observe, again and again, that the distress of a non-related but well-
known community member may elicit genuine concern in a companion, and a desire to help.

Compassion and self-sacrifice are two of the qualities we value most
in our own western society. In some cases — as when someone risks
his or her life to save another — the altruistic act is probably motivated
by the same inherent complex of helping behaviours that cause a chimpanzee to aid a companion. But there are countless instances when
the issue is clouded by cultural factors. If we know that another, especially a close relative or friend, is suffering, then we ourselves become
emotionally disturbed, sometimes to the point of anguish. Only by
helping (or trying to help) can we hope to alleviate our own distress.
Does this mean, then, that we act altruistically only to soothe our own
consciences? That our helping, in the final analysis, is but a selfish
desire to set our minds at rest? One can speculate endlessly on human
motives for helping others. Why do we send money to starving children
in the Third World? Because others will applaud and our reputation
will be enhanced? Or because starving children evoke in us a feeling
of pity which makes us uncomfortable? If our motive is to advance
our social standing, or even to alleviate our own mental discomfort,
is not our action basically selfish? Perhaps, but I feel strongly that we
should not allow reductionist arguments of this sort to detract from

the inspirational nature of many human acts of altruism. The very fact that we feel distressed by the plight of individuals we have never met, says it all.

We are, indeed, a complex and endlessly fascinating species. We carry in our genes, handed down from our distant past, deep-rooted aggressive tendencies. Our patterns of aggression are little changed from those that we see in chimpanzees. But while chimpanzees have, to some extent, an awareness of the pain which they may inflict on their victims, only we, I believe, are capable of real cruelty — the deliberate infliction of physical or mental pain on living creatures despite, or even because of, our precise understanding of the suffering involved. Only we are capable of torture. Only we, surely, are capable of evil.

But let us not forget that human love and compassion are equally deeply rooted in our primate heritage, and in this sphere too our sensibilities are of a higher order of magnitude than those of chimpanzees. Human love at its best, the ecstasy deriving from the perfect union of mind and body, leads to heights of passion, tenderness and understanding that chimpanzees cannot experience. And while chimpanzees will, indeed, respond to the immediate need of a companion in distress, even when this involves risk to themselves, only humans are capable of performing acts of self-sacrifice with *full* knowledge of the costs that may have to be borne — not only at the time, but also, perhaps, at some future date. A chimpanzee does not have the conceptual ability to become a martyr, offering his life for a cause.

Thus although our 'bad' is worse, immeasurably worse, than the worst conceivable actions of our closest living relatives, let us take comfort in the knowledge that our 'good' can be incomparably better. Moreover we have developed a sophisticated mechanism — the brain — which enables us, if we will, to control our inherited aggressive hateful tendencies. Sadly, our success in this regard is poor. Nevertheless, we should remember that we alone among the life forms of this planet are able to overcome, by conscious choice, the dictates of our biological natures. At least, this is what I believe.

And what of the chimpanzees? Are they at the end of their evolutionary progression? Or are there pressures in their forest habitat that might, given time, push them further along the path taken by our own prehistoric ancestors, producing apes that would become ever more human? It seems unlikely; evolution does not often repeat itself. Probably chimpanzees would become ever more *different* — they might, for example, develop the right side of the brain at the expense of the left.

But the question is purely academic. It could not be answered for countless thousands of years, and even *now* it is clear that the days of the great African forests are numbered. If the chimpanzees themselves survive in freedom, it will be in a few isolated patches of forest grudgingly conceded, where opportunities for genetic exchange between different social groups will be limited or impossible. And, unless we act soon, our closest relatives may soon exist only in captivity, condemned, as a species, to human bondage.

19

Our Shame

❮❮❮❮❮❮❮❮❮❮❮

EVEN THE GOMBE CHIMPANZEES are threatened by the re-
lentless march of human expansion. I was thinking about this
during a recent visit as I followed a large group of chimpanzees
high up into the open wind-swept grasslands near the crest of the rift
escarpment. I was out of breath when we arrived at our destination —
a great stand of *muhandehande* trees. As the chimps, with loud calls
of delight, began to feast on the rich crop of yellow nectar-sweet fruits,
I settled on a rock, that, shaded by one of the low, stunted trees, still
held the coolness of the night air. We were almost at the topmost peak
of the chimpanzee's world, under the pale morning sky. Below us the
ground fell away now steeply, now more gently, towards the blue-
grey expanse of Lake Tanganyika. Lines and patches of green, starting
just below the smooth golden-brown humps and ridges of the dry
upper slopes, gradually became darker and thicker, then converged as
they followed the maze of gullies and ravines that led down to the
thickly forested valleys. To the north and to the south, valley succeeded
valley, each leading its own swift-flowing stream westward, from the
watershed, high in the hills, down to the lake.

Gombe National Park, a narrow strip of rugged terrain, two miles
at its widest, stretches for no more than ten miles along the eastern
shore of the lake — a pitifully small stronghold, I reflected, for the
three communities of chimpanzees living there. For, although they still
roam free, they are effectively imprisoned — their refuge is surrounded on three sides by villages and cultivated land, while along the
fourth boundary, the shore of the lake, over one thousand fishermen

are camped. Yet these one hundred and sixty or so chimpanzees are safer than almost any other wild chimpanzees in Africa — except for those in the few remaining places, in the central part of the species' range, that are utterly remote. At least, in Gombe, there is no poaching.

I sat there, cooling down in the fresh breeze, looking out over the chimpanzees' dwindled realm. When I arrived at Gombe in 1960 you could climb to the top of the rift escarpment and gaze out to the east over chimpanzee habitat stretching into the far distance. The forests and woodlands that offered sanctuary to wildlife stretched almost unbroken from the northern tip of the lake to the southwest border of Tanzania — and beyond. There may have been as many as ten thousand chimpanzees living in Tanzania then, while today there can be no more than two thousand, five hundred. But at least many of these remaining chimpanzees are protected in two national parks, Gombe and the much larger Mahale Mountains area to the south. There are also a number of forest reserves where chimpanzees still roam in comparative safety. Chimpanzees are not eaten by any peoples in Tanzania nor has there ever been a flourishing export trade in live chimpanzees. In most other African countries where chimpanzees still live their plight is far more grim.

At the turn of the century chimpanzees were found, in their hundreds of thousands, in twenty-five African nations. From four countries they have disappeared completely. In five others, the population is so small that the species cannot long survive. In seven countries populations are less than five thousand. And even in the four remaining central strongholds chimpanzees are gradually and relentlessly losing ground to the ever-growing needs of ever-growing human populations. Forests are razed for dwellings and for cultivation. Logging and mining activities penetrate ever deeper into the natural habitats, and human diseases, to all of which chimpanzees are susceptible, follow. Moreover, the dwindling chimpanzee populations become increasingly fragmented and genetic diversity is lost until, in many cases, the small groups of survivors can no longer sustain themselves. In some countries in West and Central Africa chimpanzees are hunted for food. Even in places where they are not eaten, females are often shot, snared, chased with packs of dogs, or even poisoned in order that their infants may be captured for sale to dealers who, in turn, ship them off for the international entertainment and pharmaceutical industries, or sell them as 'pets' to anyone who will buy them.

In the tree nearest me I heard soft laughing. Fifi's two daughters, Fanni and Flossi, the edge of their appetites dulled, had begun to play. As I looked up, Fifi's most recent infant, little Faustino, reached out to touch one of the yellow fruits that his mother was chewing, then licked his fingers. Several of the chimpanzees, their hunger satiated had climbed down and were lying on the ground. Gremlin and Galahad were close to me and, even as I watched, the infant, relaxed by his mother's gentle grooming fingers, dropped off to sleep. They were five feet from where I sat and once again I was all but overwhelmed by the trust they showed, and poignantly aware of my responsibility towards them: that trust must never be broken. Galahad, dreaming perhaps, suddenly clutched his mother's hair. Gremlin responded instantly, holding him close, comforting him even as he slept so that he relaxed once more. Watching them I thought, as I so often do now, of the grim fate of hundreds of Africa's chimpanzees. Of the mothers who are killed, the infants who are seized from their arms and, shocked, terrified and hurt, dragged into a harsh and bitter new life. A life that is barren and cold because the ever-comforting arms of the mother, and the nurture and reassurance of her breast, are no more.

The whole sickening business of capturing infant chimpanzees, for any purpose whatsoever, is not only cruel but also horribly wasteful. The hunters' weapons are, for the most part, old and unreliable. Many mothers escape, wounded, only to die later of their injuries. Their infants will almost certainly die also. Often youngsters as well as their mothers are hit, particularly when the weapons are old flintlocks stuffed with nails or bits of metal. And if other chimpanzees rush to the defence of the mother and her child — then they may be shot also.

Just occasionally the hunters are thwarted. There is a true story of two hunters who set off in search of a young chimpanzee. After three days, during which they shot four mothers, three of whom escaped wounded and one who was killed along with her infant, they located and killed a fifth. She fell to the ground, her infant still living. Laying down the gun, the man went to seize the terrified, screaming infant as he clung, with the strength of desperation, to his dying mother. All at once there was a great crashing in the undergrowth and an adult male chimpanzee, hair bristling, charged out towards them. With a swiping, grabbing movement he virtually scalped one of the hunters. Seizing the other he hurled him down onto some rocks, breaking several ribs. Then he gathered up the infant and disappeared, back into the forest. When I first heard the tale I assumed that the youngster would die. But that was before we observed how Spindle looked after little Mel. Let us hope that the avenging male showed similar parental concern and skill and that the youngster was as tenacious of life as Mel. The two men managed to get to a hospital — and then, when they had recovered, were sent to jail.

Such incidents, however, must be rare. For most infants, the death of the mother brings their life in the forest to an abrupt end and leads to a succession of terrifying new experiences. After that brutal separation, the infant must first endure a nightmare journey to a native village or a dealer's camp. The captive, feet and hands often tied together with string or wire, is crammed into a tiny box or basket, or pushed into a suffocating sack. And, with each agonizing jolt, cramped and chafed by the bonds of his new captivity, freedom, comfort and joy are left further and further behind. And let us not forget that an infant chimpanzee will suffer in almost exactly the same way, emotionally and mentally, as a small human child would.

Many youngsters do not survive these journeys, for they receive little if any attention and care en route. Those that do, arrive at holding stations in a sorry plight. Many are wounded, all are dehydrated, starving and suffering from shock. Yet it is unlikely that they will find relief or solace, for the conditions that prevail at such places are typically grim and standards of care atrocious. And as they await shipment to their final destinations, still more infants will die. The survivors must then face further travel to different places around the world. At airports delays are common and there is seldom anyone to nurture the crated captives. Often, indeed, their departure is illegal so that the dealers involved, and those in their pay, do their best to conceal the existence or at least the nature of the cargo. They are the evil ones, these dealers. With the blood of countless innocents on their hands, they grow fat and rich on suffering, like those who traded human slaves in years gone by.

It is surprising that any youngsters leave their cramped air cargo crates alive, yet, against all odds, some do. Like the survivors of the concentration camps of the Third Reich, these little chimpanzees show an amazing tenacity for life. But even their arrival overseas is not necessarily the end of the road — some must travel on via tortuous routes so that their country of origin can be concealed. This is so that they can be imported as *captive*-born chimpanzees into countries that cannot legally import *wild*-born chimpanzees from Africa. And so the grim score of lives wasted continues to mount. Those youngsters that do, eventually, reach their final destinations alive, are often so weak, so emotionally damaged, that it is impossible to restore them to health. It has been estimated by those most familiar with the trade that between ten and twenty chimpanzees will die for every infant that survives to the end of the first year at its ultimate destination.

My thoughts were interrupted as the chimpanzee group, well fed and well rested, started to move down the slope. As I followed Fifi and her family my pleasure at being with them was tempered by a nagging depression. Just watching Faustino as he enjoyed the attentions of his mother and two elder sisters, constantly reminded me, after the musings of the morning, of all the unfortunate infants snatched, so abruptly, from similar family groups.

What happens to the few battered orphans that survive the horror of capture and transport? What do we offer them as reward for their endurance? Alas, only too often their lives will be so grim and wretched that it would have been better for them had they died during those bitter months when first they fell into human hands. Many infants born in captivity face an equally bleak future. The best they can hope for, these chimpanzee prisoners, is to end up in a good zoo. And zoos offering really good conditions to chimpanzees are, sad to relate, still few and far between. Because adult chimpanzees are so strong and so good at escaping, enclosures large enough to provide a proper environment are expensive. Thus countless chimpanzees languish in small cement-floored, steel-barred cells in all parts of the world. Some of these unfortunates have one or two companions with whom to share their incarceration; others must suffer alone through up to fifty years of utter boredom. They become frustrated, apathetic and, eventually, psychotic. Conditions tend to be particularly grim in many African and other Third World zoos — hardly surprising in view of the fact that hundreds of humans there must also endure deprivation and misery. But there is no excuse for the shocking conditions that still prevail in many zoos throughout Europe and in the United States.

Nor is there any excuse for the abuse of young chimps in the coastal resort areas of southern Spain and seaside areas of the Canary Islands. These youngsters, smuggled illegally into the country from Africa, are subjected to years of misery at the hands of a group of photographers who ply their trade during the holiday season, offering tourists the opportunity to be snapped holding a cute young chimpanzee dressed in children's clothes. The photos serve as reminders of a pleasant holiday in the sun — in a country that seems more exotic because of the presence of wild animals. There are, after all, no chimpanzees to be seen on the promenades of Brighton or Blackpool, or the French Riviera.

The casual tourist has no idea of the suffering inflicted on these pathetic infants. During the day they are carried or led around in the hot sun. At night many are taken into nightclubs and discos where their eyes become inflamed in smoky atmosphere and the noise must be anguish to their sensitive eardrums. Their feet are crammed into shoes that are quite the wrong shape for chimpanzee toes. They wear nappies (seldom changed) under plastic pants so that their bottoms become raw and painful. Most of them are heavily drugged. They are disciplined with blows, and some, in addition, with pressure from the tip of a glowing cigarette. As they get older their milk canines, and sometimes other teeth as well, are pulled out so that there is no risk of a customer being bitten. When they are five or six years old they are usually too big and strong for such work and they are either killed or sold to dealers.

Largely thanks to the persistent efforts of a British couple living in Spain, Simon and Peggy Templar, new legislation has been passed that enables the authorities to confiscate chimpanzees who have no permits. I was present when two of these youngsters were moved from the Templars' holding station in Spain to a sanctuary in England.

One of them, Charlie, had only been rescued a few weeks before we arrived. He was six or seven years of age. All his teeth save for three canines and the molars at the back, which were just coming through, had been knocked out. He was thin, emaciated almost. And his movements were slow and deliberate, those of an old man; he seemed knowledgeable beyond his years and weighed down by what he had learned of life. His eyes appeared to look only inwards, back at his suffering.

A British veterinarian, Kenneth Pack, who had been helping the Templars for years, was there with his blow pipe to sedate the chimpanzees so that they could be put in their travelling crates. When Charlie was hit he gazed calmly at the dart sticking from his arm, with its little tuft of red twine, then slowly pulled it out and examined it carefully. He removed the needle, then seemed to try to put it back. Finally, to my utter disbelief, he tried to inject himself. Of course he failed — there was no needle. He walked over to me and handed me the syringe. But when I made to take it from him he gently directed my hand, holding the syringe, to his arm.

The Templars had described how some of the confiscated youngsters they took in went through all the horrifying symptoms of drug withdrawal, sometimes for weeks after their arrival. As I watched Charlie, with his sad, young-old face, I was sickened. Here was an addict, trying to give himself a 'fix'.

And then there are the chimpanzees used in the entertainment industry, in circuses and movies. Of course it is possible to train chimpanzees by kindness, but the polished performances of chimpanzee stars, such as those in the various Tarzan films, *Project X*, *Bedtime for Bonzo*, and so on are, almost without exception, exacted by harsh cruelty. On the actual film sets there is rarely brutality — it would not be tolerated. But during *pre*training sessions the non-human actors-to-be are routinely beaten into submission. Often the trainer uses a lead cosh around which he rolls a newspaper. When training continues on the set, in the presence of the human actors, the rolled-up newspaper is a symbol that ensures instant obedience.

A good many captive infant chimpanzees end up as pets, particularly in Africa. Most belong to expatriates who rescue them, hunched and miserable, often close to death, from the marketplace or the roadside. Their mothers have been shot, cut up and sold for meat. There is little flesh on infants, and the hunters, if they are lucky, can get more money by selling them as pets. And so the trade continues.

At first these youngsters are easy to look after in the home. Dressed in diapers they are like living dolls, docile, affectionate, cute. They may be pampered and well cared for and when sympathetic owners take the trouble to provide a nutritious diet, security and love, the infants will enjoy their lives, unnatural though they may be. But as they grow older they become harder to manage and, by the time they are four or five years of age, they have become a nuisance and a liability. They are strong and curious. They want to explore their environment. They climb the curtains, break anything lying around, raid the refrigerator, use keys to unlock cupboards. Increasingly they must be disciplined and they resent punishment. They throw violent tantrums and bite. And so they are banished from the home, often into tiny cages on the veranda. One chimp, Socrates, had been in such a prison for months when I met him. The story of the suffering he had known in his three short years was writ clearly in his face.

Whiskey was chained. I had seen photographs of him tied up at the back of a garage, but even so I was unprepared for the surge of pure anger that swept over me when I saw him. His concrete-floored, brick-walled cell was some five by six feet. There was a gaping hole in the ramshackle roof. The tiny open cubicle was next to an Asian-type lavatory, little more than a hole in the ground with the door half open. Whiskey's 'home' had probably served the same purpose, once.

'He is like a son to me,' said the smiling Arab. I stared at him, dumbfounded. Was it stupidity or insolence that prompted this man to introduce a 'son' tied by a two-foot chain to a steel post at the back of a disused lavatory? I looked at Whiskey, met his questioning gaze. 'His chain is lengthened at night,' said his 'father'. 'Then he can move out into the garage.' Yes, I thought, at night — when chimpanzees sleep. I went up to Whiskey and he put his arms around me, returning my embrace.

As I started to move away he began hurling himself about, jerking against the chain, banging the wall with his hands, his feet. He reached out towards me, then threw a banana skin — all that he could find in his prison. Usually he threw faeces, I had been told, but everything had been cleaned in readiness for my visit.

What happens to these unfortunate chimpanzees when they become really big and strong, at adolescence? Or when their owners leave the country? Some end up in a local zoo where, even when intentions are good, funds are usually limited. Moreover, the keepers have their own families to care for, and chimpanzee fare is only too acceptable to small hungry human children. When zoos will not take the young chimps they are often killed — for most countries now have laws prohibiting their legal export. Only too often there is no haven for them in the country that is their rightful home.

There are many chimpanzees owned as pets in the United States, too. There, 'loving' owners often take steps to put off the day when they must part with them. Some chimpanzees have their teeth pulled out. One young female had both her thumbs amputated so that she couldn't (so her 'mother' thought) climb and destroy the curtains. But in the end these simian members of the family usually have to go. And by that time it is hard for them to adjust to being chimpanzees. All their lives they have been taught to behave like humans. What becomes of them, pathetic outcasts that they are? It is by no means easy to place unwanted home-raised chimpanzees in American zoos, since they tend to be socially inept and poor breeders. Often they are sold to dealers. They end up in roadside zoos, displayed in tiny cages for the ignorant to tease. Or in medical research labs.

And what is the lot of these chimpanzees used by scientists because they are so physiologically like humans? How are they treated by those who use their living bodies to try to learn more about human disease, drug addiction and mental illness? Certainly not as honoured guests in the labs. Indeed, most of them are maintained in conditions similar to those which convicts endured in bygone eras. Yet these chimpanzees are not only innocent of crime, but actually helping to alleviate human suffering. Even in the best of the labs, where breeding groups have relatively large outdoor enclosures, those chimpanzees being used in experimental procedures are housed in relatively small cages with only small outdoor enclosures. And in some of the labs that I have visited, the chimpanzees are kept in conditions that can only be described, at best, as showing absolutely no understanding of the needs of the in-mates, and at worst, as shockingly cruel.

The first laboratory I ever visited was just outside America's capital, in Rockville, Maryland. I had seen a videotape taken during an illicit entry, but even so I was not prepared for the nightmare world into which I was ushered by smiling, white-coated men. As I followed, the door to the outside closed, and all light from the sky was shut off. We moved through dimly-lit underground corridors, and I was shown room after room lined with small, bare cages, stacked one above the other, in which monkeys circled endlessly round and round, round and round. Then there was a room where young chimpanzees, two or three years old, were crammed, two together, into tiny cages mea-suring, I was told, 22 inches by 22 inches, and two feet high. They could hardly move. Not yet part of any experiment, they had already been confined there for more than three months. Those cages were enclosed in metal boxes that looked like microwave ovens — 'iso-lettes' — so that each prisoner could see out only through a small panel of glass. And what could they see? The bare wall opposite. And what was in the cage to provide occupation, comfort, stimulation? Nothing. Nothing except their own faeces and, from time to time, some food.

True, there were two chimps to a cage — at least they had each other for comfort. But not for long. Once infected — with hepatitis, AIDS or some other virulent disease — they would be separated and, like the others I saw that day, placed in cages by themselves. I watched one of these older chimpanzees, a juvenile female, as she rocked from side to side, sealed off from the outside world inside her metal isolation chamber. She was in semi-darkness. All she could hear was the inces-sant roar of air rushing through vents into her cell. When she was lifted out by one of the technicians, she sat in his arms like a rag doll, listless, apathetic. I shall be haunted forever by her eyes, and by the eyes of the other chimpanzees I saw that day. They were dull, blank, utterly without hope. Have you ever looked into the eyes of a person who, stressed beyond endurance, has given up, succumbed utterly to the crippling helplessness of despair? I once saw a little African boy whose whole family had been killed during the fighting in Burundi. He too looked out at the world unseeing, from dull, blank eyes.

Unless long-promised changes do eventually take place, there the chimpanzees will remain for the next three or four years. During that time, they will become permanently disturbed, emotionally and psychologically.

Those cages did not comply with animal welfare regulations. But even if they had, it would have made little difference. I have been saddened to find how many scientists and lab personnel see nothing wrong with the legally required minimum cage size in the United States. Hundreds of chimpanzees are confined, each one alone, in prisons measuring five feet by five feet, and seven feet high. These highly social, highly intelligent beings, whose emotions are so similar to our own, may be locked into these metal-barred boxes for life. For over fifty years.

Imagine being shut up in such a cell, with bars all around; bars on every side, bars above, bars below. And with nothing to do. Nothing to while away the monotony of the long, long days. No physical contact, ever, with another of your kind. Friendly physical contact is so terribly important for chimpanzees. Those long, relaxed sessions of social grooming matter to them, so much.

I can never forget the first time that I gazed into the eyes of a fully adult male imprisoned in one of these standard lab cages. An old motor tyre suspended from the bars above was the only object, other than himself, inside his prison. There were nine other male chimpanzees in the bleak underground room. There were no windows. Nothing to see except the other prisoners. The walls were uniform white, the doors were steel. The sounds of the chimpanzees echoed and reverberated as they greeted our arrival — myself and a veterinarian. The noise, as they hooted, and shook and beat on the bars of their prisons, was almost unbearable.

It was when they quietened that I looked into JoJo's eyes. I saw no hatred — that would have been easier to bear. Only puzzlement, gratitude that I had stopped to speak to him, to break the unbearable boredom of the day. I thought then of the chimpanzees of Gombe, free to roam the forests, free to play and groom and make nests in the springy branches. JoJo reached out a gentle finger and touched my cheek where the tears slid down into my laboratory mask.

Another nightmare visit was in Austria, just outside Vienna. To get there we drove through beautiful rolling countryside. The sun shone. In the lab the chimpanzees were locked below ground. This was a brand new building for AIDS research and anyone going into the chimpanzees quarters was required to wear heavy protective clothing. It was like struggling into an astronaut suit. If I failed to attach the nozzle of my breathing tube to the air vent in each of the rooms where I was to go, I was told, then I would suffocate. As I pulled the helmet down over my head, and felt hands zipping it closed from behind, I knew a moment of panic. My guide disappeared into a chemical shower that would sterilize his suit. I waited for the prescribed number of minutes then, peering through my glass visor, I lumbered awkwardly in his wake.

The heavy door clicked shut. In each of the three small chambers into which I was led there were two chimpanzees, each one imprisoned alone in a five-foot-square cage. There were sheets of some sort of plexiglass, or plastic, hung between the cages, through which the inmates could, I suppose, peer at each other. Most of them, I remember, just looked at us as we entered their rooms. One chimp seemed excited or afraid — I could not tell which. She came to the bars to be reassured by a clumsy, heavily gauntleted hand. As we left, they sank back into apathy — at least, no sounds followed us as the doors closed.

Throughout that brief tour around those dim, underground chambers I felt that I moved in a fantasy world, utterly remote from reality. I tried to imagine a hospital for AIDS patients — human patients — where all the doctors and nurses moved about grotesquely outfitted in space suits, where all the visitors had to strip and struggle into the same protective outfits. How terrified those chimpanzees must have been when first they saw these monstrous figures, heard the sepulchral voices distorted by the helmets. Now they are used to it. For them, the outside world, the real world with trees and sky and the comfort of normal, friendly contact with other living beings, is gone for ever.

How can the people working in these chimpanzee prisons tolerate the conditions there? Are they without feeling, without compassion? Are they utterly lacking in understanding? Are they sadistic, delighting in their power and control over such large and potentially dangerous creatures? For the most part, I think, the attitudes of the staff are forced upon them by the scientific system. Newly employed personnel are usually upset by what they see. Some quit, unable to endure the suffering around them, and feeling powerless to help. And many of those who stay on gradually come to accept the cruelty, believing (or forcing themselves to believe) that it is an inevitable part of the struggle to reduce human suffering. Some of them even become hard and callous in the process, 'all pity choked with custom of fell deed'.

Fortunately for the chimps, there are a number of compassionate people who never come to terms with the laboratory conditions, but stay because they feel that they can then try to make things better for the chimps. One such is Dr James Mahoney, who cares deeply about the 250 or so chimpanzees in his charge. It was Jim who introduced me to JoJo. And as I crouched on the floor that day, fighting my tears, Jim, who had moved off to talk to the other chimps, came back and saw my sadness. He bent down and put his arms around me. 'Don't do that, Jane,' he said. 'I have to face this every morning of my life.'

And that, of course, made the anguish worse. Jim is one of the most gentle and compassionate people I know. That insight into the hell which, for much of the time, he must endure, added a whole new dimension to my understanding. It is not only for the chimpanzees that the lab conditions must be improved — it is for the caring people, too. The technicians whose own eyes fill with unshed tears when I ask them how they can bear to supervise the snatching of infants from their mothers, the separating of a carefree youngster from the nursery for the start of his life in prison. I know that my visits bring them new hope, courage to fight on for improvements. And so, for them and for the chimpanzees, I go back again — and again. Into what, for me, is hell.

Unfortunately, those who are working, from the inside, for better conditions have a difficult and thankless task. For one thing, most of their colleagues have absolutely no understanding of *real* chimpanzee behaviour. They only know *lab* chimps. And lab chimps, deprived of almost everything they need for their physical comfort and mental stimulation, are likely to be bad-tempered, even vicious. They may spit and throw faeces, grab and bite. It is partly through frustration and aggression, partly because they are trying to establish some kind of contact with people, partly because there is almost nothing else to do. These chimps are poor ambassadors for their kind, and it is not surprising that many technicians and veterinarians dislike and even fear them.

It is true that in many labs the chimpanzees appear to be in reasonably good physical condition, despite their sterile environment. There is a mistaken belief that if animals look healthy, eat well and, above all, reproduce satisfactorily, they must be content — therefore their environment must be suitable. Change is not necessary. Of course, this is not true — certainly not where humans are concerned. Even in concentration camps, babies were born, and there is no good reason to believe that it is different for chimpanzees.

For the most part, the scientists who design the experimental conditions under which their research is to be carried out forget that they are dealing with living, sentient beings. They insist on the animals being maintained in the traditional manner. Only then, they believe, will their experiments and tests yield reliable results. A bleak, sterile and restricting environment, they say, is *necessary* for laboratory animals. Cages must be barren, without bedding or toys, because then the inmates are less likely to pick up diseases or parasites. And, of course, the cages are easier to clean when they are uncluttered. Cages must be small because otherwise it is too difficult to treat the subjects — to inject them or draw their blood. The chimpanzees must be caged individually to avoid the risk of cross-infection.

In fact, things need not be that way, and there *are* labs where more humane attitudes have led to improved conditions. Cages can be bigger because chimpanzees can be taught to approach and present their rumps for injections, their arms for the drawing of blood. They can be taught to go into smaller cages for other kinds of treatment. They can be persuaded to exchange toys, blankets and so on for food rewards so that cage cleaning is easier. And there are even a few labs where single housing of chimpanzees is the exception, rather than the rule. Recently, a number of eminent immunologists and virologists from the USA and Europe have published an article stating that experimental protocols that have, traditionally, required chimpanzees to be housed *singly* during experimentation, can, for the most part, be adapted quite satisfactorily to *pairs* of chimpanzees. This means, for all those chimpanzees currently used in hepatatis and AIDS research (the majority of all experimental animals) that the end of solitary confinement should be in sight. Surely anyone caging chimpanzees singly should be forced to prove, convincingly, to a panel of qualified scientists, the need for such inhuman conditions — particularly in view of the growing body of evidence showing that such conditions, which produce stressed animals, are not only cruel but may actually be *harmful* to the results of experiments. Because stress affects the immune system, data pertaining to drug efficiency collected from stressed subjects may be misleading.

Unfortunately, all of us who are fighting for improved conditions in the labs are up against the Establishment. And the Establishment, typically, resists change. The Establishment pits the suffering of experimental animals against the suffering of humans. Reforms, they argue, are costly. If the chimpanzees have bigger cages, social groups, an enriched environment, and better care, it will cost more. Crucial experiments will come to a grinding halt. And this, they argue, will be paid for in human suffering. This, of course, is not true. Truly essential research and testing would continue. It is difficult, on moral grounds, to justify *any* use of chimpanzees as living test tubes under even the best of conditions. That we can tolerate their continued use in lab conditions such as I have described is a damning indictment of the ethical values of our times.

In fact, the winds of change are blowing. Attitudes towards all non-human animals are changing as the general public becomes increasingly aware of the cruelty that goes on around us.

In some primate centres around the world, ethical concerns regarding the use and maintenance of our closest relatives are routinely discussed, and attempts have been or are being made to create better conditions. In some labs there are large outdoor compounds for breeding groups, and experimental animals are at least housed in pairs and given access to outdoor runs. Programmes designed to enrich the lives of the inmates are being introduced in more and more labs, to the benefit not only of the chimpanzees, but also the mental well-being of those employed to care for them. These programmes do not necessarily involve the expenditure of large sums of money — a chimpanzee's day will be far more enjoyable if he is given, for example, a magazine to read, or a comb or toothbrush and a mirror, or a tough plastic tube stuffed full of raisins and marshmallows along with a supply of twigs which he can use as tools to poke the goodies out. More sophisticated ways of alleviating boredom — such as video games — are in the planning stages.

One of the unexpected rewards I have found as I become increasingly involved in conservation and welfare issues, has been meeting so many dedicated, caring and understanding people who are fighting the same battle, fighting to improve conditions for chimpanzees in captivity, to reduce suffering, to create sanctuaries for abused or orphaned individuals, and to conserve natural habitats. These remarkable people give their time, their money — and sometimes their health — to help chimpanzees in this time of dire plight. Geza Teleki, for example, got river blindness, an incurable disease, when he worked for the government of Sierra Leone to set up a national park there specifically for chimpanzees. These people have already accomplished so much, often struggling alone against powerful adversaries. And now, as though an invisible conductor had suddenly waved his baton, many of these people are joining forces. This will, inevitably, be of great benefit to chimpanzees worldwide. (For a fuller account of the efforts to help the chimpanzees, see Appendix II.)

What, realistically, *is* the future of the chimpanzee in Africa, the wild, free and majestic being whom I have come to know so well? The best we can hope for is a series of national parks or reserves, well protected by buffer zones, where the chimpanzees and other forest denizens can live out their natural lives in peace. This, I have no doubt, we shall somehow achieve. Of course it is necessary to persuade the governments of the countries concerned that it is worth their while, that conservation of their natural resources is preferable to immediate exploitation for instant profit. Research projects bring in foreign exchange. Tourism brings in far more. The two must be planned in conjunction so that an influx of visitors disturbs neither the research nor, more importantly, the animals. Education programmes build awareness among local people. Employing field staff from villages surrounding reserved areas, as we have done at Gombe, helps the local economy and, just as important, creates enthusiasm in those people involved, enthusiasm that spreads to families and friends. This is one of the reasons why the Gombe chimpanzees are so safe from poaching.

We must remember that the people living near areas recently set aside for wildlife may have every right to feel resentful. Why should they be deprived of land that their forebears have utilized for generations past? Conservation, education and the influx of tourist dollars are not sufficient recompense. Imaginative agro-forestry projects surrounding forest reserves and parks — the growing of trees for firewood, charcoal, building poles and so on — not only protect the indigenous species, but enable people to utilize the land very much as they did in bygone years. Some conservationists tend to forget that humans are animals too!

I cannot close this chapter without sharing a story that, for me, has a truly symbolic meaning. It is about a captive chimpanzee, Old Man, who was rescued from a lab or circus when he was about eight years old and placed, with three females, on a man-made island at a zoo in Florida. He had been there for several years when a young man, Marc Cusano, was employed to care for the chimps. 'Don't go on the island,' Marc was told. 'Those brutes are vicious. They'll kill you.'

For a while Marc obeyed instructions, and threw the chimps their food from a little boat. But soon he realized that he could not care for them properly unless he established some kind of rapport with them. He began going ever closer and closer when he fed them. One day Old Man reached out and took a banana from Marc's hand. How well I remember when, at Gombe, David Greybeard first took a banana from mine. And, as for me with David, that was the start of a relationship of mutual trust between Marc and Old Man. Some weeks later Marc actually went on to the island. Eventually he could groom and even play with Old Man, although the females, one of whom had a baby, were more standoffish.

One day as Marc was cleaning up the island he slipped and fell. This startled the infant, who screamed, and his mother, her protective instinct aroused, at once leaped to attack Marc. She bit his neck as he lay, face down, on the ground, and he felt the blood run down his chest. The other two females rushed to support their friend. One bit his wrist, the other his leg. He had been attacked before, but never with such ferocity. He thought it was all up for him.

And then Old Man charged to the rescue of this, his first human friend in years. He dragged each of the highly roused females off Marc and hurled them away. Then he stayed, close by, keeping them at bay, while Marc slowly dragged himself to the boat and safety. 'Old Man saved my life, you know,' Marc told me later, when he was out of hospital.

If a chimpanzee — one, moreover, who has been abused by humans — can reach out across the species barrier to help a human friend in need, then surely we, with our deeper capacity for compassion and understanding, can reach out to help the chimpanzees who need us, so desperately, today. Can't we?

20

Conclusion

⋘⋘⋘⋘

IT IS THIRTY YEARS since I began to study chimpanzees. Thirty years during which there has been much change in the world, including the way in which we think about animals and the environment. My own personal journeys during this period, through the peaceful forests of Gombe and through the thorny jungles that have sprung up around issues of animal welfare and conservation, have led me a very long way from the naive young English girl who, with her mother, stepped so eagerly from the boat onto the Gombe beach. Yet she is still there, still part of the more mature me, whispering excitedly in my ear whenever I see some new or fascinating piece of chimpanzee behaviour — not only at Gombe, but sometimes in a captive situation also. I am as thrilled today, when I first see a new baby close up, when a mother reaches, with a slight pout of concern, to gather up her straying child, when one of the big males charges past, hair bristling, lips compressed in magnificent pride, as I was during the earliest months of the study.

My journeyings among the chimpanzees have been enriched by experiences more exciting and rewarding than any we could have imagined, back at the start of it all. The harvest — the understanding that has come from long hours spent with our closest living relatives — has opened many windows onto a world all but unknown thirty years ago. How fortunate for me that fate directed my footsteps to Louis Leakey and he, in turn, directed them to Tanzania — where, for all these years, I have been able to pursue the quest for more and ever more knowledge, helped and supported by one of the most stable, peaceful and conservation-conscious governments in all of Africa.

The information gathered at Gombe, along with that from other study sites in Africa and from research on captive chimpanzees, has enabled us to paint a fascinating portrait of our closest living relative, an ever more detailed likeness of a highly complex being. Of course the picture is not yet completed — we have neither plumbed the depths of chimpanzee aggressiveness, nor measured the upper limits of their capacity for care and compassion. We have not been studying them for long enough — after all, thirty years represents but two-thirds of the chimpanzee's life span. Above all our experience at Gombe has emphasized the need for long-term study if we are to attempt to understand the complex society of these chimpanzees. Much of their social behaviour only began to make sense when we had been among them long enough to work out who was related to whom among the adults. And only by staying there, year after year, were we able to document the close, supportive and enduring bonds that grow up between family members. Moreover, had the research come to an end after a mere ten years we should never have observed the brutality that can occur during intercommunity clashes. If it had stopped after twenty years we should not have documented the touching story of little Mel's adoption by adolescent Spindle. And who knows what the next decade will reveal? That there will be more surprises I do not doubt, for every year, from 1960 onward, has brought its own rewards in terms of new observations about chimpanzee nature, new insights into the workings of their mind. They are such complex beings, their behaviour so flexible, their individuality so pronounced.

Over the years we have become gradually more and more familiar with an ever growing number of chimpanzees, each with his or her own vivid and unique personality. What a rich cast of characters, each one moulded by the complex interplay of genetic inheritance and experience, family life and the historical era into which he or she was born. For chimpanzees, like humans, have their history. Epidemics of polio and pneumonia, and a series of violent intercommunity interactions not dissimilar to human warfare, have ravaged their community. There were the dark years when Passion and Pom, infant killers, cannibals, made it unsafe for mothers and their newborn babies to walk through the seeming peace of the forest. There have been struggles for power every bit as dramatic as those surrounding the successions of human kings and dictators. And I have been privileged, since the early sixties, to record these facts — to compile the history of a group of beings who have no written language of their own.

As in human societies, certain individuals have played key roles in shaping the fortunes of their community. Some of the adult males who have demonstrated outstanding leadership qualities of determination, courage or intelligence would figure prominently in chimpanzee history books: Goliath Braveheart, Mike of the Cans, Brutal Humphrey, Figan the Great, Goblin the Tempestuous. There would be epic accounts of how they strove for power and won. And other individuals have played major roles, also. But for Hugh and Charlie the Kasakela community might never have divided. Without Gigi and the gatherings of roused, excited males she has always attracted, her community might well have been less aggressive, less martial in its attitude to neighbours.

But the community males were strong, their victories impressive. Imagine, if the chimpanzees could talk, the stirring tales that would be told around the fire of the Four Years War against the Kahama deserters, the liquidation of the rebel males who turned their backs on their long-time friends and tried to make it on their own. And what stories, too, would be woven around the repelling of the Kalande and Mitumba invaders when — it was rumoured — Humphrey and Sherry lost their lives in defence of the realm. And how the females would love to sing the praises of Gigi, living legend, Amazon dowager of her community.

The bizarre behaviour of Passion, infamous murderess, and her daughter Pom, would be analysed in all the criminal literature. And mothers would threaten their naughty children: 'Passion will get you if you don't behave'.

They would have their myths too, the chimpanzees. They would honour the wise ones of old who first taught them how to open the ground and fashion tools for the capture of ants and termites, and how to intimidate their enemies with rocks and clubs. And the adolescents would learn how to propitiate the great god Pan, sylvan deity of all wild creatures, with impressive waterfall ceremonies and rain dances deep in the heart of the forest.

And of course there would be a myth concerning White Ape who so suddenly appeared in their midst. Who was greeted initially with fear and anger, but whose coming led, eventually, to the provision of bananas — magical, like the dropping of manna from heaven. David Greybeard would figure in the legend, too — the one chimpanzee who had no fear of White Ape and introduced her to the forest world of his kind.

In fact, if Louis Leakey had not sent me to Gombe in 1960 the chimpanzees would almost certainly have lost their refuge, for there was, at that time, a move afoot among the local inhabitants to change the reserved status of the area so that they could move back in and cultivate the land. But the interest my study aroused around the world ensured Gombe's continuing protected status. The chimpanzees, knowing this, would naturally have made me their patron saint!

How, in actual fact, *do* they perceive me? Me and the other humans who have moved in to watch them and shared in the documentation of their history? Today, I believe, we are taken for granted. In the chimpanzee's scheme of things, other chimpanzees are the most important figures, particularly close family and friends — and the current alpha male. Animals, such as monkeys, bushpigs and so forth, are important too as a source of food. Baboons, often ignored, are also regarded as potential competitors for precious resources, except for young baboons who are perceived by young chimpanzees as potential playmates. And humans, at Gombe, are regarded simply as another animal species, as a natural component of the chimpanzee's environment. Unthreatening, occasional providers of bananas. Sometimes irritating since they tend to be noisy in the undergrowth, but for the most part benign and harmless.

Of course, the chimpanzees recognize us as individuals. Many of them are more relaxed when I am with them than they are in the presence of other human observers. This, I believe, is because I invariably follow them quite alone, and also because I remain quietly in the background, intruding as little as possible, often foregoing op-

portunities to collect additional data, or getting a photo of some particular behaviour, if this means disturbing or irritating the chimpanzees I am with. For the most part the chimpanzees are very tolerant
also of the Tanzanian field staff, the men who work with them day
in, day out, month after month, year after year. But they are usually
ill at ease if they encounter strange Africans in the park. I have been
with chimps who, hearing a group of fishermen moving along one of
the paths from lake-shore to village, have crouched, still and silent, in
the bushes or long grass until the men passed. A few of the chimps
avoid tourists — indeed, the shier females no longer visit camp unless
they are part of a big group in which case, clearly, they believe there
is security in numbers. But some, particularly those who grew up during the days of heavy student involvement, actually appear to find
tourists, and all their odd — and unsuitable — costumes, of some interest. At least, that is what it seems when Fifi or Gigi or Prof move
close to a camera-clicking, sun-burned group and lounge nearby,
grooming each other — or just sitting.

The nature of my own relationship with the chimpanzees is, to some
extent, constrained by our research methods at Gombe. We deliberately keep our distance from the chimps, partly because they are much
stronger than us and can be dangerous if they lose their respect for
humans, partly so as to influence their natural behaviour as little as
possible. We do try to administer medicine if a chimp is sick or hurt,
but for the most part we simply observe and record. The chimps are
in no way dependent on me, not even for bananas which they often
receive very irregularly indeed. This is probably why I do not, as many
suppose, think of the chimps as extensions of my own family. I have
the most profound regard and respect for them. I am endlessly fascinated by their behaviour and I can spend hours, days in their company. Often I am asked if I prefer chimpanzees to humans. The answer
to that is easy — I prefer some chimpanzees to some humans, some
humans to some chimpanzees! Because, of course, they are all so different. One or two whom I have known, like Humphrey and Passion,
I disliked very much indeed. Others, like David Greybeard and Flo,
Gilka, Fifi and Gremlin, have a very real hold on my heart, and my
affection for them is close to love. But it is a love for beings who are

essentially wild and free. And because I do not groom or play with them, or take part in their disputes, it is a one-sided love — they do not love me back, as does a child or a dog. This in no way diminishes what I feel for them.

I shall never forget sitting by Flo's dead body and, some ten years later, below the nest where Melissa breathed her last. As I thought back over their lives, I knew a real sense of loss, and I mourned their passing as I have grieved at the passing of close human friends. When little Getty was found dead, his body mutilated, I was numbed by shock and horror, and again I felt deep sorrow. No longer would I be able to watch his exuberant play, record his innovative games, delight in his fearless, adventurous spirit.

Of all the Gombe chimpanzees, though, it is David Greybeard whom I have loved the most. His body was never found. He simply stopped coming to camp and, as the weeks became months, we gradually realized that we would never see him again. Then I felt a sorrow deeper than that which I have felt for any other chimpanzee, before or since. I am glad I was spared the anguish I should have known had I seen him, too, in death. David Greybeard, gentle yet determined, calm and unafraid, David Greybeard who opened my first window onto the chimpanzee's world.

And what a magic world this is for me, a world far removed from the bustle of modern society, where I can find peace, and energy. A world with power to heal the battered spirit. For in the forest there is a sense of timelessness, and in the lives of the chimpanzees, so like us, so different, a quality that brings one face to face with basic realities. They get on with living, and, although things can go very wrong sometimes, for the most part they enjoy that living to the full.

It was to Gombe that I went, seeking solace, after Derek lost his heroic battle with cancer. He died in Germany, where, for a while, we had hope for a miracle cure — a hope that we clutched at, desperately, as do thousands of others in similar circumstances. When hope was ended, I knew that bitterness and despair that comes to all of us when we lose one whom we have loved. I spent a little time with my family in England. Then back to Dar, with all its sad associations: gazing each day at the Indian Ocean where Derek, despite his crippled legs, had found freedom swimming among his beloved coral reefs. It was a real relief to leave the house and bury myself, for a while, in Gombe. For there I could hide my hurt among the ancient trees, find new strength for living in the forests that, surely, have changed little since Christ walked the hills of Jerusalem.

It was during that time, when I spent hours in the field with little thought of collecting data, that I came closer to the chimpanzees than ever before. For I was with them not to observe, to learn, but simply because I needed their company, undemanding and free of pity. And, as my spirit gradually healed, so I became increasingly aware of a new intuitive empathy with the chimpanzees, with these closest living relatives of ours. Ever since, I have felt more in tune with the natural world, the endless cycles of nature, the interdependence of all living things in the forest.

I shall never, so long as I live, forget one afternoon that I spent in the company of Fifi and her family and Evered. For three hours I followed as the chimpanzees, peaceful and harmonious, wandered from place to place, now feeding, now resting and grooming while the youngsters played. Towards the end of the afternoon they moved down into the Kakombe Valley and, following the Kakombe Stream eastward, headed for the fig trees — *Mtobogolo* the local people call them — that grow near the Kakombe waterfall. As we drew near, the roar of falling water sounded ever louder in the soft green air. Evered and Freud, hair bristling, moved faster. Suddenly the waterfall came into sight through the trees, cascading down from the stream bed fifty feet or more above. Over countless aeons the water has worn a deep groove in the sheer rock. On either side lianas hang, looping down the rock face. Vivid green ferns wave ceaselessly in the wind created by the rushing of the water through its rocky channel.

All at once Evered charged forward, leapt up to seize one of the hanging vines, and swung out over the stream in the spray-drenched wind. A moment later Freud joined him. The two leapt from one liana to the next, swinging into space, until it seemed the slender stems must snap or be torn from their lofty moorings. Frodo charged along the edge of the stream, hurling rock after rock now ahead, now to the side, his coat glistening with spray.

For ten minutes the three performed their wild displays while Fifi and her younger offspring watched from one of the tall fig trees by the stream. Were the chimpanzees expressing feelings of awe such as those which, in early man, surely gave rise to primitive religions, worship of the elements? Worship of the mystery of water, which seems alive; always rushing on, yet never going; always the same, yet ever different.

The ritual over, the chimpanzees turned from the stream and climbed into the fig tree where Fifi sat. They all began to feed on the ripe fruits with grunts of pleasure. A gentle breeze rustled the branches and the little stars of light that shone through the dancing canopy above, gleamed and winked. Pervading everything was the almost intoxicating scent of ripe figs, the humming of insects, the chirps and flutters and whirring flight of feeding birds. The huge branches of the fig tree were festooned with vines, twining and twisting up towards the sky. Their flowers gave nectar to butterflies, to iridescent sun birds. The chimpanzees munched figs, spitting out the seeds so that new figs would grow. The tree, one day, would crash to earth with all its burden of plant and animal life, and from the decaying richness a host of new life would spring forth. Everywhere life entwined with life, uniting with death to perpetuate the forest home of the chimpanzees. An endless cycle, ancient as the first trees. Old patterns repeated in ways that would always be new.

In the richness of such a lush environment lived the chimpanzee-like creatures that became the first men. Slowly they evolved. Some became more adventurous and left the forest on excursions into the surrounding savanna, in search of food or new territory. What a relief it must have been after the danger of such ventures, to return to the safety of the forest. But gradually, just as earlier life forms became increasingly independent of seas and lakes and rivers, so did early humans learn to live away from the forest. They found caves and fire, learned to build dwellings, to hunt with weapons, to talk. And then they became bold and arrogant. They began to hack at the outskirts of the forest itself, bending to their will that which for so long had nurtured them. Today, striding the face of the globe, humans clear the trees, lay waste the land, cover mile upon mile of rich earth with concrete. Humans tame the wilderness and plunder its riches. We believe ourselves all-powerful. But it is not so.

Relentlessly the desert inches forward, gradually replacing the life-sustaining forests with barren and uncompromising harshness. Plant and animal species vanish, lost to the world before we have learned their value, their place in the great scheme of things. World temperatures soar, the ozone layer is depleted. All around we see destruction and pollution, war and misery, maimed bodies and distorted minds, human and non-human alike. If we allow this desecration to continue we shall, ourselves, be doomed. We cannot meddle so greatly in the master plan and hope to survive.

Thinking of this whole terrible picture, the magnitude of our sin against nature, against our fellow creatures, I was overwhelmed. How could I — or anyone — make a difference in the face of such vast and mindless destruction?

A fig dropped close by, startling me. Fifi climbed from the tree and lay near me with closed eyes, replete. Here, at least, was perfect trust between humans and animals, perfect harmony between creatures and their wild environment. Faustino, tottering a little, moved close to me and, with his wide-eyed stare, reached to touch my hand, then wandered back to Fifi. Trust. And freedom. I thought of the countless chimpanzees who have lost their forest homes, and of the prisoners in zoos and labs around the world. I remembered the story of Old Man and how he had responded to the need of a human friend.

The will to fight, to fight to the bitter end, flared up. The chimpanzees need help now more than ever before, and we can only help if we each do our bit, no matter how small it may seem. If we don't, we are betraying not only the chimpanzees but also our own humanity. And we must never forget that, insurmountable as the environmental problems facing the world may seem, if we all pull together we have a good chance of bringing about change. We must. It is as simple as that!

Evered, Freud and Frodo climbed down and, with Fifi and Faustino, moved away, deep into the peace of the forest. I watched them go, then looked back. And where the sun shone through a window in the dense vegetation, a rainbow had appeared, spanning the spray-cloud at the foot of the waterfall.

Some Thoughts on the Exploitation of Non-Human Animals

The more we learn of the true nature of non-human animals, especially those with complex brains and correspondingly complex social behaviour, the more ethical concerns are raised regarding their use in the service of man — whether this be in entertainment, as 'pets', for food, in research laboratories or any of the other uses to which we subject them. This concern is sharpened when the usage in question leads to intense physical or mental suffering — as is so often true with regard to vivisection.

Biomedical research involving the use of living animals began in an era when the man in the street, while believing that animals felt pain (and other emotions) was not, for the most part, much concerned by their suffering. Subsequently, scientists were much influenced by the Behaviourists, a school of psychologists which maintained that animals were little more than machines, incapable of feeling pain or any human-like feelings or emotions. Thus it was not considered important, or even necessary, to cater to the wants and needs of experimental animals. There was, at that time, no understanding of the effect of stress on the endocrine and nervous systems, no inkling of the fact that the use of a stressed animal could affect the results of an experiment. Thus the conditions in which animals were kept — size and furnishings of cage, solitary versus social confinement — were designed to make the life of the caretaker and experimenter as easy as possible. The smaller the cage the cheaper it was to make, the easier to clean, and the simpler the task of handling its inmate. Thus it was hardly surprising that research animals were kept in tiny sterile cages, stacked one on top of the other, usually one animal per cage. And ethical concern for the animal subjects was kept firmly outside the (locked) doors.

As time went on, the use of non-human animals in the laboratories increased, particularly as certain kinds of clinical research and testing on *human* animals became, for ethical reasons, more difficult to carry out legally. Animal research was increasingly perceived, by scientists and the general public, as being crucial to all medical progress. Today it is, by and large, taken for granted — the accepted way of gaining new knowledge about disease, its treatment and prevention. And, too, the accepted way of testing all manner of products, destined for human use, before they go on the market.

At the same time, thanks to a growing number of studies into the nature and mechanisms of animals' perceptions and intelligence, most people now believe that all except the most primitive of non-human animals experience pain, and that the 'higher' animals have emotions similar to the human emotions that we label pleasure or sadness, fear or despair. How is it, then, that scientists, at least when they put on their white coats and close the lab doors behind them, can continue to treat experimental animals as mere 'things'? How can we, the citizens of civilized, western countries, tolerate laboratories which — from the point of view of the animal inmates — are not unlike concentration camps? I think it is mainly because most people, even in these enlightened times, have little idea as to what goes on behind the closed doors of the laboratories, down in the basements. And even those who do have some knowledge, or those who are disturbed by the reports of cruelty that are occasionally released by animal rights organizations, believe that *all* animal research is essential to human health and progress in medicine and that the suffering so often involved is a *necessary* part of the research.

This is not true. Sadly, while some research is undertaken with a clearly defined objective that could lead to a medical breakthrough, a good many projects, some of which cause extreme suffering to the animals used, are of absolutely no value to human (or animal) health. Additionally, many experiments simply duplicate previous experiments. Finally, some research is carried out for the sake of gaining knowledge for its own sake. And while this is one of our more sophisticated intellectual abilities, should we be pursuing this goal at the expense of other living beings whom, unfortunately for them, we are able to dominate and control? Is it not an arrogant assumption that we have the *right* to (for example) cut up, probe, inject, drug and implant electrodes into animals of all species simply in our attempt to learn more about what makes them tick? Or what effect certain chemicals might have on them? And so on.

We may agree that the general public is largely ignorant of what is going on in the labs, and the reasons behind the research there, rather as the German people were mostly uninformed about the Nazi concentration camps. But what about the animal technicians, the veterinarians and the research scientists, those who are actually working in the labs and who know exactly what is going on? Are all those who use living animals as part of standard laboratory apparatus, heartless monsters?

Of course not. There may be some — there are occasional sadists in all walks of life. But they must be in the minority. The problem, as I see it, lies in the way we train young people in our society. They are victims of a kind of brainwashing that starts, only too often, in school and is intensified, in all but a few pioneering colleges and universities, throughout higher science education courses. By and large, students are taught that it is ethically acceptable to perpetrate, in the name of science, what, from the point of view of the animals, would certainly qualify as torture. They are encouraged to suppress their natural empathy for animals, and persuaded that animal pain and feelings are utterly different from our own — if, indeed, they exist at all. By the time they arrive in the labs these young people have been programmed to accept the suffering around them. And it is only too easy for them to justify this suffering on the grounds that the work being done is for the good of humanity. For the good of the one animal species which has evolved a sophisticated capacity for empathy, compassion and understanding, attributes which we proudly acclaim as the hallmarks of humanity.

I have been described as a 'rabid anti-vivisectionist'. But my own mother is alive today because her clogged aortic valve was removed and replaced by that of a pig. The valve in question — a 'bioplasticized' one, apparently — came, we were told, from a commercially slaughtered hog. In other words, the pig would have died anyway. This, however, does not eliminate my feelings of concern for that particular pig — I have always had a special fondness for pigs. The suffering of laboratory pigs and those who are raised in intensive farming units has become a special concern of mine. I am writing a book, *An Anthology of the Pig*, which, I hope, will help to raise public awareness regarding the plight of these intelligent animals.

Of course I should like to see the lab cages standing empty. So would every caring, compassionate human, including most of those who work with animals in biomedical research. But if all use of animals in the laboratory was *abruptly* stopped there would probably, for a while anyway, be a great deal of confusion, and many lines of enquiry would be brought to a sudden halt. This would inevitably lead to an increase in human suffering. This means that, until alternatives to the use of live animals in the research labs are widely available and, moreover, researchers and drug companies are legally compelled to use them, society will demand — and accept — the continued abuse of animals on its behalf.

Already, in many fields of research and testing, the growing concern for animal suffering has led to major advances in the development of techniques such as tissue culture, *in vitro* testing, computer simulation and so on. The day will eventually come when it will no longer be necessary to use animals at all. It must. But much more pressure should be brought to bear for the speedy development of additional techniques. We should put far more money into the research, and give due acknowledgement and acclaim to those who make new breakthroughs — at the very least a series of Nobel prizes. It is necessary to attract the brightest in the field. Moreover, steps should be taken to insist on the use of techniques already developed and proven. In the meantime, it is imperative that the numbers of animals used be reduced drastically. Unnecessary duplication of research must be avoided. There should be more stringent rules regarding what animals may and may not be used for. They should be used only for the most pressing projects that have clear-cut health benefits for many people, and contribute significantly to the alleviation of human suffering. Other uses of animals in the labs should be stopped *immediately*, including the testing of cosmetics and household products. Finally, so long as animals are used in our labs, for any reason whatsoever, they should be given the most humane treatment possible, and the best possible living conditions.

Why is it that only relatively few scientists are prepared to back those who are insisting on better, more humane conditions for laboratory animals? The usual answer is that changes of this sort would cost so much that all progress in medical knowledge would come to an end. This is not true. Essential research would continue — the cost of building new cages and instigating better care-giving programmes would be considerable, but negligible, I am assured, when compared with the cost of the sophisticated equipment used by research scientists today. Unfortunate, though, many projects are poorly conceived and often totally unnecessary. They might indeed suffer if the costs of maintaining the research animals are increased. People making their living from them would lose their jobs.

When people complain about the cost of introducing humane living conditions, my response is: 'Look at your life-style, your house, your car, your clothes. Think of the administrative buildings in which you work, your salary, your expenses, the holidays you take. And, after thinking about those things, *then* tell me that we should begrudge the extra dollars spent in making a little less grim the lives of the animals used to reduce human suffering.'

Surely it should be a matter of moral responsibility that we humans, differing from other animals mainly by virtue of our more highly developed intellect and, with it, our greater capacity for understanding and compassion, ensure that medical progress speedily detaches its roots from the manure of non-human animal suffering and despair. Particularly when this involves the servitude of our closest relatives.

In the United States, federal law still requires that every batch of hepatitis B vaccine be tested on a chimpanzee before it is released for human use. In addition, chimpanzees are still used in some highly inappropriate research — such as the effect on them of certain addictive drugs. There are no chimpanzees in the labs in Britain — British scientists use chimpanzees in the United States, or at the TNO Primate Centre in Holland where EEC funding has recently gone into a new chimpanzee facility. (British scientists do, of course, make massive use of other non-human primates and thousands of dogs, cats, rodents and so forth.)

The chimpanzee is more like us than is any other living being. Physiological similarities have been enthusiastically described by scientists for many years, and have led to the use of chimpanzees as 'models' for the study of certain infectious diseases to which most non-human animals are resistant. There are, of course, equally striking similarities between humans and chimpanzees in the anatomy of the brain and nervous system, and — although many have been reluctant to admit to these — in social behaviour, cognition and emotionality. Because chimpanzees show intellectual abilities once thought unique to our own species, the line between humans and the rest of the animal kingdom, once thought to be so clear, has become blurred. Chimpanzees bridge the gap between 'us' and 'them'.

Let us hope that this new understanding of the chimpanzees' place in nature will bring some relief to the hundreds who presently live out their lives as prisoners, in bondage to Man. Let us hope that our knowledge of their capacity for affection and enjoyment and fun, for fear and sadness and suffering, will lead us to treat them with the same compassion that we would show towards fellow humans. Let us hope that while medical science continues to use chimpanzees for painful or psychologically distressing experiments, we shall have the honesty to label such research for what, from the chimpanzees point of view, it certainly is — the infliction of torture on innocent victims.

And let us hope that our understanding of the chimpanzee will lead also to a better understanding of the nature of other non-human animals, a new attitude towards the other species with which we share this planet. For, as Albert Schweitzer said, 'We need a boundless ethic that includes animals too.' And at the present time our ethic, where non-human animals are concerned, is limited and confused.

If we, in the western world, see a peasant beating an emaciated old donkey, forcing it to pull an oversize load, almost beyond its strength, we are shocked and outraged. That is cruelty. But taking an infant chimpanzee from his mother's arms, locking him into the bleak world of the laboratory, injecting him with human diseases — this, if done in the name of Science, is not regarded as cruelty. Yet in the final analysis, both donkey and chimpanzee are being exploited and misused for the benefit of humans. Why is one any more cruel than the other? Only because science has come to be venerated, and because scientists are assumed to be acting for the good of mankind, while the peasant is selfishly punishing a poor animal for his own gain. In fact, much animal research is self-serving too — many experiments are designed in order to keep the grant money coming in.

And let us not forget that we, in the west, incarcerate millions of domestic animals in intensive farm units in order to turn vegetable protein into animal protein for the table. While this is usually excused on grounds of economic necessity, or even regarded by some as sound animal husbandry, it is just as cruel as the beating of the donkey, the imprisonment of the chimpanzee. So are the fur farms. So is the abandonment of pets. And the illegal puppy farms. And fox hunting. And much that goes on behind the scenes when animals are trained to perform for our entertainment. The list could get very long.

Often I am asked whether I do not feel that it is unethical to devote time to the welfare of 'animals' when so many human beings are suffering. Would it not be more appropriate to help starving children, battered wives, the homeless? Fortunately, there are hundreds of people addressing their considerable talents, humanitarian principles and fund-raising abilities to such causes. My own particular energies are not needed there. Cruelty is surely the very worst of human sins. To fight cruelty, in any shape or form — whether it be towards other human beings or non-human beings — brings us into direct conflict with that unfortunate streak of *inhumanity* that lurks in all of us. If only we could overcome cruelty with compassion we should be well on the way to creating a new and boundless ethic — one that would respect all living beings. We should stand at the threshold of a new era in human evolution — the realization, at last, of our most unique quality: humanity.

Chimpanzee Conservation
and Sanctuaries

Throughout the western world, and in many Third World countries, attitudes towards animals and the environment are changing. There is more awareness of the plight of chimpanzees than there was a few years ago, and with it a growing concern and desire to help. In answer to special needs, individuals emerge when they are most wanted.

Deeply involved in trying to instigate and assist conservation strategies in Africa is the Committee for Conservation and Care of Chimpanzees — 'The Four C's'. This is a body of scientists, all of whom are concerned with chimpanzee conservation and welfare. Its chairman is Dr Geza Teleki, who is working with Dr Toshisada Nishida and others to put together an action plan designed to help as quickly as possible the beleaguered chimpanzees across the African continent. The map on the next page shows the places where chimpanzees are still found, and the research projects, some of which (such as those at Gombe and Mahale Mountains in Tanzania, Tai Forest in Ivory Coast, and Lope in the Gabon) have been in progress for a good many years. In all cases these projects are highly beneficial to chimpanzee conservation in the immediate vicinity.

Surveys are needed desperately in many countries, to find out more about the actual range of chimpanzees today. And, in certain key areas, it is important to set up research projects as quickly as possible. Without such projects, carried out in conjunction with conservation education, tourism and agro-forestry, chimpanzees will disappear rapidly from a number of other countries. Of course, the studies will be important in their own right, too. They will enable us to learn more about one of the most fascinating aspects of chimpanzee behaviour — that about which we know least — behavioural differences between populations in different parts of Africa. As it is, not only

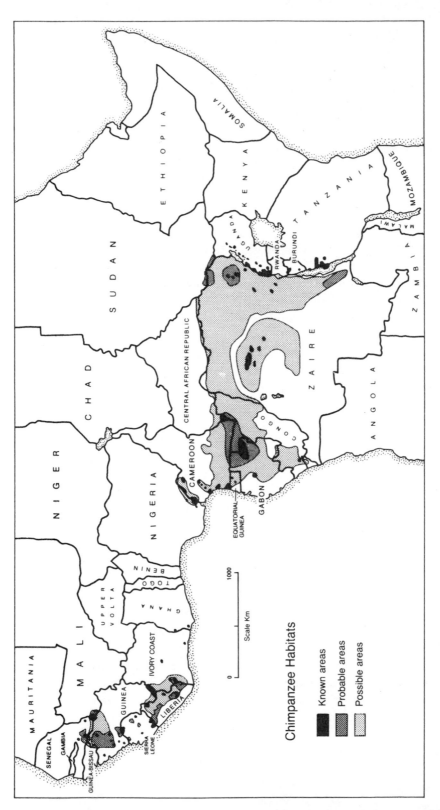

The distribution of chimpanzees in Africa. The main concentrations of chimps remaining in Africa are in those countries with the largest undisturbed tracts of forests, such as Zaire, Gabon and Cameroon. (Map reproduced courtesy of Dr. Geza Teleki and the Committee for Conservation and Care of Chimpanzees)

Chimpanzee Habitats

Known areas

Probable areas

Possible areas

are hundreds of individual chimpanzees perishing but, in addition, whole cultures are vanishing before we have had time to study them.

During 1989 I became involved in chimpanzee conservation and protection in Burundi, one hundred miles or so north of Gombe along Lake Tanganyika. This was a direct result of the conservation interests of Ambassador James D. Phillips (Dan) and his wife, Lucie. I first visited Burundi at their invitation, met with President Buyoya and a number of his ministers and other members of government, including the Secretary General, Venant Bambonehoyo, and was sincerely impressed by the efforts being made by this regime to save the remaining forested areas of their beautiful country. I was impressed, too, by the steps towards chimpanzee conservation that had already been taken. I met with Peter Trenchard, coordinator of the Biological Diversity Project, who had spent several months observing the chimpanzees of the Kibira National Park, a lovely mountain rain forest in the north of the country. I was taken by Paul Cowles and Wendy Bromley to visit a small group of chimpanzees in the south of the country. There a number of local people had been employed as 'Chimp Guards' to monitor the movements of the chimpanzees as they travel from one strip of gallery forest to another, crossing cultivated areas and bypassing native villages. The close juxtaposition of chimpanzees and villagers is not unusual; the steps being taken to preserve the chimpanzees — initially set up by a conservationist of great foresight, Robert Clausen — I found unique. But the situation was potentially explosive since the farmers living nearby need land badly. Paul (who had worked first as a Peace Corps Volunteer and was then the Catholic Relief Services technical consultant to the National Institute for the Environment and Nature Conservation (INECN)) explained the agro-forestry project with which he was involved. First nurseries are developed for fast-growing tree species. The seedlings are then planted around villages. Many of the trees can be used within two years — for building poles, charcoal, firewood, as shade trees, and for enriching the soil with nitrogen. Each tree species has its own special function. The application of this project for the protection of the remaining indigenous forested areas is obvious. Wendy was working with Paul, explaining this new concept to the villagers. Burundi is to be congratulated on this programme, without which it might be impossible to conserve wild chimpanzees in this very small country that has such a high human population density.

In order to provide additional income and incentive to local people, it is clearly necessary to develop controlled tourism. As a first step, Charlotte Uhlenbroek, funded by the Jane Goodall Institute (UK), began to habituate a group of chimpanzees in the south of the country to the presence of humans. As an integral part of this programme (which is, of course, aimed at collecting as much data on the behaviour of the chimpanzees as possible) a number of the Chimp Guards have already visited Gombe for training in observational methods by the Tanzanian field staff.

A new awareness and interest in chimpanzees in the country brought to light the fact that there were a number of pet chimpanzees in the capital, Bujumbura, and in other places throughout the country. Most of these young-sters have almost certainly been smuggled across the border from neigh-bouring Zaire. Thanks to the support of the government and the help of many individuals, JGI (UK), in close cooperation with INECN, is now able to go ahead with the building of a sanctuary near Bujumbura where ex-pets, and any confiscated youngsters, can be released. This sanctuary has been planned, a site has been allocated and, with the help of Steve Matthews, the initial construction will commence during 1990. The first two orphans, Poco and Socrates, are in a temporary cage in the garden of Melinda (Mimi) Brian. An education centre, where local people and visitors can learn about chim-panzees and their behaviour, will be an important and central part of the sanctuary.

In the same year Karen Pack set out for Pointe Noire in Congo-Brazzaville to try to set up a sanctuary for ex-pet chimpanzees and those confiscated from hunters by the government. Karen is currently working for JGI (UK) at the zoo in Pointe Noire to enrich the environment of eight chimpanzees there. These eight will, it is hoped, join a number of ex-pets and confiscated youngsters in a sanctuary to be constructed by the JGI. An education centre is planned, along the same lines as the one in Burundi. This will be with the full support and approval of the Congolese government. Once again, Steve Matthews will mastermind the construction, with the generous support of Conoco Inc. — an oil company that is showing genuine care for the envi-ronment. We are especially grateful to Roger Simpson. Until the sanctuary is built, Madame Jamart is caring for the young chimpanzees confiscated by the government. She and her husband are remarkable.

These will certainly not be the first sanctuaries for abused or unwanted chimpanzees. The first in Africa was begun in the late 1960s by Eddie Brewer. As a government official in charge of wildlife, Eddie confiscated young chim-panzees being smuggled through the Gambia (where, by then, chimpanzees had become extinct). His daughter, Stella, subsequently moved the chimps to Senegal, where every effort was made to reintroduce them into a natural habitat. Unfortunately, the wild chimpanzees objected to the intrusion into their territory, and it was necessary to remove the ex-captives and relocate them on Baboon Island on the Gambia River. For many years now this project has been carried on by an extraordinarily dedicated individual, Janice Carter.

A truly remarkable British couple living in Zambia, Sheila and David Siddle, have turned their home into a refuge for confiscated youngsters. Chimpanzees are not endemic to Zambia, and most of the orphans were confiscated after being smuggled out of Zaire. The Siddles have constructed a remarkable eight-acre enclosure and have an ambitious plan to fence in a huge area of bushland where, eventually, the whole group can roam in relative freedom. The new Liberian Animal Orphanage and Rehabilitation Centre has an enclosure for chimpanzees and there are plans afoot for the development of additional sanctuaries in Zaire and Kenya. In Uganda, confiscated youngsters at the Entebbe Zoo are badly in need of enlarged quarters. In almost every country in Africa where chimpanzees still live there is this problem of orphans. The exception is Tanzania, where, I can proudly announce, there are only two ex-pets (from Zaire) who will soon, we hope, find final refuge with the Siddles.

In Chapter 19 I introduced those champions of abused chimpanzees, Simon and Peggy Templar. Some of their confiscated youngsters went out to the Gambia, but more recently the battered orphans from the Spanish beach racket have been finding refuge at Monkey World in Dorset, England. This sanctuary was created through the combined efforts of Jim Cronin, Steve Matthews and veterinarian Ken Pack. Some of these youngsters were in pitiful shape when they arrived, but they were nurtured, played with, disciplined and loved by Jeremy Keeling, a truly caring person whose exceptional rapport with chimpanzees has done much to heal their mental scars.

Wallace Swett began a most remarkable facility, Primarily Primates, in Texas, USA. There, among about twenty other chimpanzees, is Virgil (whose real name is Willie), star of the movie *Project X*, along with his 'girlfriend' Ginger (real name Harry). They are living with the most extraordinarily varied collection of once abused chimpanzees from all over the United States.

Two biomedical labs have piloted 'retirement' schemes for chimpanzees not currently being used in experiments. Fred Prince, of the New York Blood Center, has put a number of ex-lab chimpanzees on some small islands in Liberia. Jorg Eichburg, of the Southwest Biomedical Foundation, has built more conventional cages with outdoor runs. Neither is a place where I would want to spend my declining years, were I a chimp, but anything is better than a small lab cage. And the *concept* is a major step in the right direction.

In summary, the plight of chimpanzees worldwide is grim. In Africa there is desperate need for funding — for surveys, for studies and for sanctuaries — and a need too for qualified and dedicated people to conduct the surveys and studies, and to work with confiscated or abandoned orphan chimpanzees. There is a growing need for sanctuaries outside Africa too, as illegal shipments of chimpanzees are confiscated in various countries, as individuals are rescued from the entertainment and pet trade, and as others are retired from laboratory research. Yet somehow I feel sure that wonderful and dedicated people like those who have already contributed so much and provided so many homeless chimpanzees with sanctuary — and love — will appear. Humans, through ignorance and greed, have brought hundreds of chimpanzees to this pitiful state; humans, through concern and compassion, must do what they can to address the wrong.

Note. Information about all these people and places can be obtained from the Jane Goodall Institute for Research, Education and Conservation, P.O. Box 26846, Tucson, Arizona 85726; from the Jane Goodall Institute (UK), 10 Durley Chine Road South, Bournemouth BM2 5HZ; or from the Jane Goodall Institute (Canada), P.O. Box 3125, Station 'C', Ottawa, Ontario K1Y 4J4.

Acknowledgments

How, after almost thirty years, can I begin to thank adequately all the people who have made it possible to continue the research at Gombe? It is hard, looking back, to differentiate between contributions towards the actual study and contributions to my personal well-being. After all, the years at Gombe, the observing and documenting of the lives of the chimpanzees, are so entwined with my own personal life that it is often difficult to separate the two. Possibly I should not even try. But this means that I should write a whole separate book, for the help and support that I have been given has been so great. Sometimes I have been overwhelmed by the kindness, generosity and desire to help that I have found in people around the world. It makes for a warmth around the heart that has, again and again, given me strength to cope when times were rough.

I believe, and hope, that I did express my gratitude to all those who helped Gombe during the first ten years of the study — in my first book, *In the Shadow of Man*. Here I will try to express my thanks to those people and organizations who have enabled me to carry on since then.

First I must mention my gratitude to the Tanzanian government: to our past President, Mwalimu Julius Nyerere, now Chairman of the Party, conserver of forest habitats, a botanist in his own right and to his successor, President Hassan Mwinyi, and to the many individuals in various governmental departments who have been so helpful and supportive over so many years. Special thanks to the various regional commissioners and district development directors of the Kigoma Region who have given assistance at all times, and to the Director of Wildlife, Costa Mlay. Very special thanks to the Director of Tanzania National Parks, David Babu, and many of the Wardens; to the Director of the Tanzania Wildlife Research Institute, Karim Hirji; to the Director and staff (especially Addie Lyaruu) of the Tanzania Scientific Research Council.

Many foundations, institutions and individuals have contributed funds over the past twenty years. For the National Geographic Society a very special word of gratitude. The Society funded the entire research programme for many years and continues to support the work in a variety of ways. The publicity that the Gombe chimpanzees have received over the years through magazine articles, TV programmes and, more recently, magazine advertisements, has, more than any other single factor, made it possible for me, and for all those helping me, to raise money for various chimpanzee programmes. I must mention, especially, Melvin Payne, Gil Grosvenor, Mary Smith and Neva Folk who, over recent years, have been so very, very helpful.

The LSB Leakey Foundation made a number of generous grants, and especial thanks to Tita Caldwell, Gordon Getty, George Jagels, Coleman Morton, and Debbie Spies for their support and friendship.

Many individuals have made donations that have helped to maintain the research at Gombe since the generous funding of the Grant Foundation ended soon after the 1975 kidnapping, when forty armed men abducted four students (this is described in Chapter 7). The donors are too numerous to mention by name, but my heartfelt thanks go to every one of you, not only for major contributions, but also for the smaller gifts that represent the same magnanimous spirit on the part of the givers. One of my most treasured donations came to me in Africa from a small boy who mailed a quarter, taped to a sheet of paper, with a note saying that there would be more when he could earn money himself.

And let me thank so much my good friend Jim Caillouette who has long assisted with medical supplies for the Tanzanian staff.

A number of companies have made donations, and I should especially like to thank Jeff Walters and the Sony Corporation for generously providing a number of video cameras, players and tapes for the recording of behaviour in the field.

Many people have been helpful in Kigoma, nearest town to Gombe. Special thanks to Blanche and Toni Bescia, Subhadra and Ramji Dharsi, Rahma and Christopher Liundi, Asgar Remtulla and Kirit and Jayant Vaitha.

I am continually grateful to Robert Hinde for his patience with me in the early days when I was his student, and for his subsequent help and support. And my thanks also to David Hamburg who, in 1972, negotiated an affiliation between Gombe and Stanford University and secured major funding that enabled a succession of talented students to work at Gombe as research assistants, giving new vigour to the project.

I cannot mention by name, here, all the students and assistants who took part in observing the chimpanzees and collecting data. But I am specially grateful for the major contributions of those who remained in the field for several years: Harold Bauer, David Bygott, Patrick McGinnis, Larry Goldman, Hetty and Frans Plooij, Anne Pusey, Alice Sorem Ford, Geza Teleki, Mitzi Thorndal, Caroline Tutin and Richard Wrangham. Also Curt Busse and David Riss, who did the fifty-day follow of Figan.

Next I come to the Tanzanian Field Assistants, for whose skilled work and dedication I have the highest respect. These men have worked at Gombe for many years — the work is their life. After the 1975 kidnapping our research would have come to an end had it not been for these men. My special thanks go to Hilali Matama who commenced work at Gombe in 1968 and is now in day-to-day charge of the data collections, and to Hamisi Mkono and Eslom Mpongo who have also been with me for over ten years. I thank also Yahaya Alamasi, Ramadhani Fadhili, Bruno Helmani, Hamisi Matama, Gabo Paulo. I would like to pay very special tribute to the late Mzee Rashidi Kikwale who passed away in 1988. Rashidi it was who accompanied me on my early travels through the mountains of Gombe. With him I saw my first chimps. Throughout the subsequent years Rashidi was a loyal worker and a real friend. Towards the end of his life he played an important role at Gombe, acting as honorary headman of staff camp. After he died, Hilali lamented: 'We are like a body without a head.' He is greatly missed.

Two other people who have made major contributions to the research at Gombe are Christopher Boehm and Anthony Collins. Chris introduced the use of 8mm video cameras into the day-to-day recording equipment and trained certain of the Tanzanian field staff to use them. This has resulted in capturing, on tape, many unique incidents. And I am able to *see* much of what has gone on when I have been away. Tony is Field Director in charge of the baboon study. During his twice-yearly three-month visits he also takes over most of the administration — for the hours he has put into working out problems (wages, benefits, insurance and so on), as well as for his dedication to the baboon research, I am eternally grateful. More recently British veterinarian Kenneth Pack has entered the scene. For his opportune visit, just in time to help save the life of that very special chimpanzee, Goblin, I shall always be immensely grateful. And also for his skilled treatment of the Gombe baboons when disease recently swept through the study troops.

A wonderful team of people has helped with data analysis and administration in Dar es Salaam. For eight years Trusha Pandit was my right-hand woman — there was nothing she could not turn her hand to. She has recently left to go with her husband to India, and no one will ever quite replace her.

Other people who have spent hour upon hour helping with data analysis and all manner of other aspects of running Gombe (and organizing me!) are Jeanee Deane, Jenny Gould, Jennifer Hanay, Ann Hinks, Uta Soutter and Judy Taylor. My heartfelt thanks to all. And, too, to those wonderful friends who sprang into the breach after Derek's death, helping and strengthening me in many ways: first and foremost, of course, members of my own warm and supportive family, Vanne (who was to undergo open-heart surgery herself just a few months later), Olly, Audrey and Judy. And Grub — poor fellow, having a mother always immersed in chimpanzees and chimpanzee talk. In Dar es Salaam there was Derek's son, Ian. And Clarissa and Gunar Barnes, Jenny and Michael Gould, Frauke and Benno Haffner, Sigy and Ted McMahon, Nancy and Robert Nooter, Trusha and her husband Prashant, Judy and Adrian Taylor. And very special friends, with whom I stayed during the first miserable days after my return to Tanzania, Dick Viets and his wonderful wife, Marina who, tragically, passed away recently and is missed so deeply, but also remembered with much love and affection, by so many. And others, who have been of great help subsequently: Liz and Ron Fennell, Catherine and Tony Marsh, Penelope Breeze and Stevenson McIllvaine, Mollie and David Miller, Julie and Don Petterson, and Dimitri Mantheakis and his sons.

  Next I must try to thank those who made possible the Jane Goodall Institute for Research, Conservation and Education, a tax-exempt organization into which all donations are now channelled. It was conceived by the late Prince Ranier di san Faustino and his wife, Genevieve. After his death, Genie worked hard and made his dream a reality with the help of other wonderful friends — Joan Cathcart, Bart Deamer, Margaret Gruter, Douglas Schwartz, Dick Slottow and Bruce Wolfe. So much effort, so much generosity in time, or money, or both. Subsequently other loyal supporters have been part of the Institute: Larry Barker, Ed Bass, Hugh Caldwell, the late Sheldon Campbell, Bob Fry, Warren Iliff, Jerry Lowenstein, Jeff Short, Mary Smith and Betsy Strode. Here a very special word of thanks to two people whose generosity did much to set the Institute on its feet — Gordon and Ann Getty whose fabulous challenge gift in 1984 gave us, for the first time, an endowment. And my heartfelt thanks, too, to William Clement who made extraordinarily generous contributions when the Institute moved from San Francisco to Tucson, Arizona. I must also express gratitude to staff members who have worked so hard for so little to help me realize some of my dreams over the past few years. To Sue Engel, for helping the fledgling Institute to fly. And to Jennifer Kenyon and ChimpanZoo coordinator, Virginia Landau. A number of people have generously donated their effort and time, and I specially thank Leslie Groff, Gale Paulin, and Humphrey and Penny Taylor. And how can I properly express my thanks to Robert Edison and Judy Johnson who have led the effort to build up the Institute over the past few years. Bob, in particular, shares all my values where animal welfare is concerned. Next, I must express my thanks to Geza Teleki who, after fighting for chimpanzee conservation and welfare virtually single-handed subsequent to his return from Sierra Leone, has now joined forces with the JGI. Geza, in fact, is 'Our Man in Washington', where he heads the Committee for Conservation and Care of Chimpanzees (The Four C's). Geza, along with Heather McGiffin, also provides wonderful hospitality every time I visit America's capital — which, these days, is many times a year. Other people who have been deeply involved in efforts to improve things for chimpanzees, and who have been very helpful in Washington, are Michael Bean, Bonnie Brown, Roger Coras, Kathleen Mozzoco, Senator John Melcher, Ron Nowak, Nancy Reynolds, Christine Stevens and Elizabeth Wilson.

Many others have made great contributions, each in his or her own unique way, and to all of them I am so grateful — especially to Michael Aisner for great fund-raising efforts and true dedication to the cause; to Mark Maglio for contributing terrific art work; and to Peggy Detmer, Trent Meyer and Bart Walter for their wonderful bronzes.

More recently still the Jane Goodall Institute (UK) was born. Already this is a strong organization — because of the remarkable people who agreed to join the Board as Trustees: Robin Brown, Mark Collins, Genevieve di san Faustino, Robert Hinde, Bertil Jernberg, Guy Parsons, Victoria Pleydell-Bouverie, Sir Laurens van der Post, Susan Pretzlik, Karsten Schmidt, John Tandy, Steve Matthews, the late Sir Peter Scott — and my mother, Vanne. Along with Karsten Schmidt who steered the Institute safely through the Charitable Trust Commission, the bulk of the day-to-day work is done by Guy Parsons, Robert and Dilys Vass, Steve Matthews, Sue Pretzlik, and Vanne. The success of the launching of this Institute was due also to the generous donation by the Condor Preservation Trust, arranged by Robin and Jane Cole, much hard work by Clive Hollands and his staff, and the contributions, in terms of books and posters, by Michael Neugebauer. We are off to an auspicious start and hope to do much in Britain to raise awareness about the plight of chimpanzees, particularly among children. And many people, such as John Eastwood, Pat Groves, Neil Margerison, and Pippit Waters, are out there helping us.

It is difficult to express adequately my indebtedness to my late husband, Derek Bryceson, for his help, support and advice. Without him I doubt I could have kept the research going after the kidnapping of 1975. Derek, with his vast knowledge and understanding of Tanzania, helped me to train the field staff and to reorganize the data collection. Many were the discussions I had with him on puzzling aspects of chimpanzee behaviour; his comments, proffered from the point of view of a farmer, were often penetrating, and gave me new insights. His contribution was indeed great; even now, because he was so loved and honoured in Tanzania, his name confers on me, his widow, a position I would never otherwise have attained.

Now I must try to thank my mother, Vanne, for the staggering contribution she has made. Not only did she encourage my childhood dream of studying wild animals, but, of course, she even accompanied me to Gombe in 1960. Her wisdom and advice over the sometimes stormy years between then and now have been invaluable. She has helped with fund raising, she has read and commented on manuscripts, and she has always been a tower of strength. And, of course, there would have been no book without her — for I should not have been!

Finally, there are the chimpanzees themselves, all those unique, vivid personalities: Flo and Fifi, Gilka and Gigi, Melissa and Gremlin, Goliath and Mike, Figan and Goblin, Jomeo and Evered. And David Greybeard who, although he moved on to the Happy Hunting Grounds over twenty years ago, remains closest to my heart.

Index

The letters B, C and CG indicate that the animal named is a baboon, a chimpanzee, or a chimpanzee at Gombe.